MATHEMATICS RESEARCH DEVELOPMENTS

EMERGING CONCEPTS IN EVOLUTION EQUATIONS

MATHEMATICS RESEARCH DEVELOPMENTS

Additional books in this series can be found on Nova's website under the Series tab.

Additional e-books in this series can be found on Nova's website under the e-book tab.

MATHEMATICS RESEARCH DEVELOPMENTS

EMERGING CONCEPTS IN EVOLUTION EQUATIONS

CAROLYN MURPHY
EDITOR

NOTICE TO THE READER

Library of Congress Cataloging-in-Publication Data

Names: Murphy, Carolyn (Writer on mathematics), editor.
Title: Emerging concepts in evolution equations / Carolyn Murphy, editor.
Description: Hauppauge, New York: Nova Science Publishers, Inc., [2017] | Series: Mathematics research developments | Includes bibliographical references and index.
Identifiers: LCCN 2017000721 (print) | LCCN 2017001444 (ebook) | ISBN 9781536108613 (softcover) | ISBN 9781536108736
Subjects: LCSH: Evolution equations. | Evolution equations, Nonlinear. | Differential equations, Partial.
Classification: LCC QA377.3 .E44 2017 (print) | LCC QA377.3 (ebook) | DDC 515/.35--dc23
LC record available at https://lccn.loc.gov/2017000721

Published by Nova Science Publishers, Inc. †New York

CONTENTS

PREFACE

This book reviews new research and analyzes emerging concepts in evolution equations. Chapter One discusses the evolution equation of Lie-type for finite deformations, and its time-discrete integration. Chapter Two presents a review of recent results on group analysis of nonlinear evolution equations in one spatial variable. Chapter Three addresses the problem of exponential stabilization of a class of 1-D PDEs with Dirichlet boundary control.

As explained in Chapter 1, the theory of evolution equations of Lie-type analyses a class of systems of time-dependent first-order ordinary differential equations on a Lie group (resp. homogeneous space), which are generated by vector fields related to a corresponding finite dimensional Lie algebra. Their interesting geometric features give rise to important tools, and have originated new mathematical techniques and notions used for investigating differential equations. In Chapter 1, the authors shall identify such a type of evolution equation within solid mechanics wherein it describes the evolution of finite deformations on the space of all symmetric positive-definite matrices resp. on the general linear group. In fact, while the position and shape of a deformed body take place in the usual three-dimensional Euclidean space $\mathbb{R}^3$, a corresponding progress of the deformation tensor $\mathbf{C}$ makes up a trajectory in the space of all symmetric positive-definite matrices – a negatively curved Riemannian symmetric manifold (a specific homogeneous space). The authors prove that a well-known relation between deformation rate $\partial\mathbf{C}$ and symmetric velocity gradient $\mathbf{d}$, via deformation gradient $\mathbf{F}$, can actually be interpreted as an equation of

Lie-type describing evolution of the deformation tensor **C** on the configuration space. The same applies to deformation gradient **F**, which evolves on the general linear group.

As a consequence, this identification leads to geometrically consistent time-discrete integration schemes for finite deformation processes, such as the Runge–Kutta–Munthe-Kaas method, or also briefly mentioned the semi-discrete Magnus and Fer expansion methods.

In Chapter 2 the authors present a review of our recent results on group analysis of nonlinear evolution equations in one spatial variable. The main emphasis is on the novel systematic approaches that expand the range of applications of the traditional Lie approach to evolution equations. The authors analyze two possible ways of expanding the applicability of Lie group methods: considering more general types of symmetries and applying the group approach to more general (fractional) evolution equations.

In Chapter 3, the authors address the problem of exponential stabilization of a class of 1-D PDEs with Dirichlet boundary control. The authors first reformulate the control problem in semigroups context on a suitable state space. They also convert the initial system to an auxiliary one by the use of a Volterra transformation with a kernel satisfying an appropriate PDE. Secondly, Chapter 3 proves the exponential stabilization of the target system and we show some useful spectral properties. Thirdly, the authors deal with the exponential stability of the initial system. In semigroup context, this corresponds to construct a boundary controller which causes an exponential decay of the semigroup $S(\cdot)$ representing the closed-loop system. Precisely, $\|S(t)\| = O(e^{-\lambda t})$, $t \to \infty$, for some $\lambda > 0$. The proposed method provides an explicit solution of the Lyapunov's equation by means of the kernel PDE and the parameters of the system in consideration. Further, the technics of constructing the feedback law offer operational information about the stabilization mechanism.

In: Emerging Concepts in Evolution Equations ISBN: 978-1-53610-861-3
Editor: Carolyn Murphy, pp. 1-30

Chapter 1

The Evolution Equation of Lie-Type for Finite Deformations and Its Time-Discrete Integration

Zdeněk Fiala*

Institute of Theoretical and Applied Mechanics CAS,
Prague, Czech Republic

Abstract

The theory of evolution equations of Lie-type analyses a class of systems of time-dependent first-order ordinary differential equations on a Lie group (resp. homogeneous space), which are generated by vector fields related to a corresponding finite dimensional Lie algebra. Their interesting geometric features give rise to important tools, and have originated new mathematical techniques and notions used for investigating differential equations. Here, we shall identify such a type of evolution equation within solid mechanics wherein it describes the evolution of finite deformations on the space of all symmetric positive-definite matrices resp. on the general linear group. In fact, while the position and shape of a deformed body take place in the usual three-dimensional Euclidean space $\mathbb{R}^3$, a corresponding progress of the deformation tensor $\mathbf{C}$ makes up a trajectory in the space of all symmetric positive-definite matrices – a negatively curved Riemannian symmetric manifold (a specific homogeneous space). We prove that a well-known relation between deformation

*E-mail address: fiala@itam.cas.cz.

rate $\partial\mathbf{C}$ and symmetric velocity gradient $\mathbf{d}$, via deformation gradient $\mathbf{F}$, can actually be interpreted as an equation of Lie-type describing evolution of the deformation tensor $\mathbf{C}$ on the configuration space. The same applies to deformation gradient $\mathbf{F}$, which evolves on the general linear group.

As a consequence, this identification leads to geometrically consistent time-discrete integration schemes for finite deformation processes, such as the Runge–Kutta–Munthe-Kaas method, or also briefly mentioned the semi-discrete Magnus and Fer expansion methods.

Keywords: evolution equation, Lie group integrators, solid mechanics, kinematics of finite deformation, symmetric positive-definite matrices, Runge–Kutta–Munthe-Kaas method, Magnus expansion method, Fer expansion method

AMS Subject Classification: 74A05, 34C40, 65L06

1. Introduction

The theory of *evolution equations of Lie-type* analyses a class of systems of time-dependent first-order ordinary differential equations on a Lie group G (resp. homogeneous space M)

$$\frac{d}{dt}m_t = \psi_*(a_t)(m_t) \qquad a : \mathbb{R}^+ \to \mathfrak{g}\,, \tag{1}$$

which are generated by vector fields $\psi_*(a_t)$ related to a corresponding finite dimensional Lie algebra $\mathfrak{g}$ of G via (3). Evolution equations of Lie-type are otherwise known as *Lie systems*. Their interesting geometric features give rise to important notions and tools, which have originated new mathematical techniques used for investigating differential equations.

To be specific, let the Lie group G act on the manifold M *transitively* by

$$\Psi : G \times M \to M, \tag{2}$$

where transitivity means that any two points $n, m \in M$ can be related by some group element $g \in G$ via $n = \Psi(g, m)$. The operation Ψ then induces at the infinitesimal level an action

$$\psi_* : \mathfrak{g} \times M \to \mathfrak{X}(M) \tag{3}$$

from the Lie algebra $\mathfrak{g}$ ($\equiv T_eG-$ the tangent space to G at the identity element e) into the Lie algebra of all vector fields $\mathfrak{X}(M)$ on M. The resulting *fundamental vector fields* ξ_a indexed by Lie algebra elements are defined by

$$\xi_a(m) = \psi_*(a)(m) \equiv T_e\Psi^m(a) := \frac{\partial}{\partial s}\bigg|_{s=0} \Psi^m(\gamma(s)) \in T_mM \tag{4}$$

in terms of a curve $\gamma(s)$ in G, for which $\gamma(0) = e$ and $\gamma'(0) = a \in \mathfrak{g}$. Denoting by $\Psi^m \equiv \Psi(., m)$, then $T_e\Psi^m$ stands for the corresponding linear *tangent map*, which to any Lie algebra element a assigns a particular vector $\xi_a(m)$ at the point $m = \Psi^m(e) \in M$ from the corresponding tangent space T_mM. Fundamental vector fields thus at any point of the manifold represent directions of possible motions by means of the action of the group, and so are also known as the infinitesimal generators of this group action. For more information about the Lie algebra of vector fields on manifolds in general see [19], and those related to our purposes see [9, 17] in particular.

One of the main properties of Lie systems is that their solutions admit a non-linear superposition principle allowing us to write the general solution in terms of any fundamental set of particular solutions [5]. However, our main interests will concern the numerical analysis– namely a particular class of geometric numerical integrators [15], whose great virtues derive from the fact that the local structure in a neighborhood of any point of a Lie group can be described by its Lie algebra, which then enables to adapt classical numerical procedures [7, 17, 21].

In this contribution, whose exposition draws on [12] which was primarily aimed at engineering community, we shall identify this type of evolution equations within solid mechanics wherein it describes the evolution of finite deformations. Actually, while the position and shape of a deformed body take place in the usual three-dimensional Euclidean space $\mathbb{E}^3$, a corresponding progress of the deformation tensor $\mathbf{C}_t$ makes up a trajectory in the space of all symmetric positive-definite matrices $Sym^+(3, \mathbb{R})$– a negatively curved Riemannian symmetric manifold. We prove that a well-known classical relation [13, 24]

$$\frac{\partial}{\partial t}\mathbf{C}_t = 2\mathbf{F}^T\mathbf{d}\mathbf{F} \tag{5}$$

between deformation rate $\partial\mathbf{C}_t/\partial t$ and symmetric velocity gradient $\mathbf{d}$, via deformation gradient $\mathbf{F}$, can be actually interpreted as an *equation of Lie-type* describing evolution of the right Cauchy – Green deformation tensor $\mathbf{C}_t$ on the configuration space $Sym^+(3,\mathbb{R})$. Due to an additional dependence of $\mathbf{C}_t(X)$ on the point of reference configuration X as well, see (6) and (11), we make use of partial derivation in time. As a consequence, this interpretation leads to geometrically consistent time-discrete integration schemes for finite deformation processes, such as the Runge – Kutta – Munthe-Kaas method, or the briefly mentioned semi-discrete Magnus and Fer expansion methods.

2. Kinematics of Finite Deformation

Let us consider a continuous body [13, 24], which before deformation occupies a region $\mathcal{B}$ of the three-dimensional Euclidean point space $\mathbb{E}^3$ considered now as the Riemannian manifold $\mathcal{E}^3$. The region is assumed open and bounded, with smooth boundary. The body $\mathcal{B}$ represents a reference configuration made up of points labeled by capital letters X. By symbol $\mathcal{S} = \Phi(\mathcal{B}) \subset \mathcal{E}^3$ we denote a deformed or actual configuration, which is placed in the Riemannian manifold $\mathcal{E}^3$– the ambient space whose points are labelled by small letters x.

Globally, deformation is represented by a diffeomorphism $\Phi : \mathcal{B} \to \mathcal{E}^3$, and deformation process by a one-parameter group of diffeomorphisms $\Phi : I \times \mathcal{B} \to \mathcal{E}^3$ indexed by time. Within solid mechanics, however, one adopts a local point of view to describe a deformation process in terms of time-dependent deformation tensor fields. To be specific, a deformation gradient

$$\mathbf{F} := \frac{\partial \Phi}{\partial X} \equiv T_X\Phi : T_X\mathcal{B} \to T_x\mathcal{S}, \quad \text{where } x = \Phi(X) \tag{6}$$

– actually a tangent map to the deformation $\Phi : \mathcal{B} \to \mathcal{S} = \Phi(\mathcal{B})$ between tangent spaces $T_X\mathcal{B}$ and $T_x\mathcal{S}$ is defined as

$$(T_X\Phi)(v) := \left.\frac{\partial}{\partial s}\right|_{s=0} \Phi(\gamma(s)), \tag{7}$$

where $\gamma(s)$ is a curve in $\mathcal{B}$ such that $X = \gamma(0)$ and $v = \gamma'(0)$. We use

$$\Phi_* \equiv T\Phi \equiv \mathbf{F} \qquad \text{push-forward} \tag{8}$$

$$\Phi^* = (\Phi_*)^{-1} = \mathbf{F}^{-1} \qquad \text{pull-back} \tag{9}$$

as a common terminology, where operations push-forward Φ_* and pull-back Φ^* extend their applicability to tensors of any order, see [13, 24] and cf. (10), (11), (29).

Now, if we denote by $\mathbf{g}$ a Riemannian metric on the actual configuration $\mathcal{S} \subset \mathcal{E}^3$, and similarly by $\mathbf{G}$ a Riemannian metric on the referential configuration $\mathcal{B}$, then the covariant *right Cauchy-Green deformation tensors* [13, 24]

$$\mathbf{C}^\flat \equiv \mathbf{GC} = \Phi^*(\mathbf{g}) \tag{10}$$

is an image of $\mathbf{g}$ from actual configuration to referential one induced by deformation Φ. It describes local geometry of deformed body $\mathcal{S}$ from the point of view of an observer attached to the undeformed body $\mathcal{B}$. That is, $\mathbf{C}^\flat$ is a Riemannian metric on $\mathcal{B}$. Since its mixed form

$$\mathbf{C} = \mathbf{F}^T\mathbf{F} \in Sym^+(3, \mathbb{R}) \tag{11}$$

can be expressed in terms of deformation gradient $\mathbf{F}$ and its transpose, $\mathbf{C}$ is a symmetric positive-definite matrix.

When considering a deformation process Φ_t, the corresponding time-indexed right Cauchy – Green deformation tensors form a curve $\mathbf{C}_t$ in $Sym^+ \equiv Sym^+(3, \mathbb{R})$, with the deformation rate [13, 24]

$$\partial\mathbf{C}_t := \frac{\partial}{\partial t}\mathbf{C}_t = 2\mathbf{F}^T\mathbf{dF} \in sym(3, \mathbb{R}). \tag{12}$$

Both $\partial\mathbf{C}_t$ and the rate-of-deformation tensor (stretching) $\mathbf{d}$– the symmetric velocity gradient $\mathbf{L} := \nabla v$ (via covariant derivative), are simply symmetric matrices from the vector space $sym \equiv sym(3, \mathbb{R})$.

Since only strictly positive coefficients in linear combinations are allowed, Sym^+ forms in sym an open symmetric cone, and so a manifold [10]. Actually, Sym^+– the space of all right Cauchy – Green deformation tensors at a material point $X \in \mathcal{B}$ is a negatively curved Riemannian (globally) symmetric manifold [1, 2, 4, 20, 28]. Due to its openness in sym, the tangent space $T_{\mathbf{C}_0}Sym^+$ to Sym^+ at a point $\mathbf{C}_0$ is isomorphic to sym

$$T_{\mathbf{C}_0}Sym^+(3, \mathbb{R}) \simeq sym(3, \mathbb{R}). \tag{13}$$

Note that this tangent space is formed by tangent vectors $\partial\mathbf{C}_0$ to all curves $\mathbf{C}_t$ – deformation processes, passing through the point $\mathbf{C}_0$. From the viewpoint of the manifold Sym^+, deformation rates $\partial\mathbf{C}$ are thus *vectors* (contravariant vectors), whereas deformation tensors $\mathbf{C}$ are *points*.

For the deformation gradient $\mathbf{F}$, a similar equation to the relation (12) applies [13, 24]

$$\partial\mathbf{F}_t := \frac{\partial}{\partial t}\mathbf{F}_t = \mathbf{LF}_t\,, \tag{14}$$

now on the general linear group of all non-singular matrices $GL(3, \mathbb{R})$.

2.1. Lie Group of Transformations

The Lie group of transformations [19, 25] enters the configuration space Sym^+ naturally via a composition of successive deformations, making Sym^+ a *homogeneous space*. In fact, if we consider two successive deformations Φ_1 and Φ_{21} resulting in the single one Φ_2, together with corresponding deformation gradients,

$$\Phi_2 = \Phi_{21} \circ \Phi_1 \qquad \mathbf{F}_2 = \mathbf{F}_{21}\mathbf{F}_1 , \tag{15}$$

then

$$\mathbf{C}_2 = \mathbf{F}_1^T \mathbf{C}_{21} \mathbf{F}_1 := R(\mathbf{F}_1, \mathbf{C}_{21}). \tag{16}$$

That is, the resulting Cauchy – Green deformation tensor $\mathbf{C}_2$ is obtained by "translating" deformation tensor $\mathbf{C}_{21} := \mathbf{F}_{21}^T \mathbf{F}_{21}$ – the relative deformation of state 2 with respect to state 1, by an "amount" $\mathbf{F}_1$.

Deformation gradients with matrix multiplication (15) form the general linear group of all non-singular matrices $GL \equiv GL(3, \mathbb{R})$. Due to the exchange of order in composition $R(\mathbf{F}_1, R(\mathbf{F}_2, .)) = R(\mathbf{F}_2\mathbf{F}_1, .)$, compared to the order of matrix multiplication in GL, the operation

$$R : GL \times Sym^+ \to Sym^+ \qquad (\mathbf{F}, \mathbf{C}) \mapsto \mathbf{F}^T\mathbf{C}\mathbf{F} \tag{17}$$

is called the *right translation*. This group operation is *transitive*, that is, for any two deformation tensors $\mathbf{C}_1, \mathbf{C}_2 \in Sym^+$, there is a group element $\mathbf{F}_{21} \in GL$, such that $\mathbf{C}_2 = R(\mathbf{F}_{21}, \mathbf{C}_1)$, see for example [19].

2.2. Configuration Space $Sym^+(N)$ as a Homogeneous Space

Since deformation gradients $\mathbf{F}$ belong to the general linear group $GL(N) \equiv GL(N, \mathbb{R})$ with matrix multiplication as the group operation, where $N = 3$, there is a natural map

$$p : GL \to Sym^+ \qquad \mathbf{F} \mapsto \mathbf{C} := \mathbf{F}^T\mathbf{F}. \tag{18}$$

As many elements of $GL(N)$ may have in $Sym^+(N)$ the same image, the map does not have an inverse. However, taking into account polar decomposition of a nonsingular matrix $\mathbf{F} = \mathbf{R}\mathbf{U}$, where $\mathbf{R} \in O(N) \equiv O(N, \mathbb{R})$ – the *orthogonal group*, and $\mathbf{U} \in Sym^+(N)$, we see that all elements in the *right coset* $[\mathbf{U}] := \{\mathbf{R}\mathbf{U} \mid \mathbf{R} \in O(N)\}$ will have the same image $\mathbf{U}^2 \in Sym^+(N)$. As a result, we get a factorization of $GL(N)$ into disjoint right cosets, and so an identification of the *homogeneous space* $Sym^+(N)$ with the right coset space

$$Sym^+(N,\mathbb{R}) \simeq GL(N,\mathbb{R})/O(N,\mathbb{R}) \equiv GL/O. \tag{19}$$

The correspondence

$$\hat{p}: GL/O \to Sym^+ \qquad [\mathbf{F}] \mapsto \mathbf{C} := \mathbf{G}^T\mathbf{G} \quad \text{for some } \mathbf{G} \in [\mathbf{F}] \tag{20}$$

is one-to-one, and in addition, it is a diffeomorphism– an invertible and differentiable mapping in both directions.

Operation $\rho_{\mathbf{G}}(.) := \mu(. , \mathbf{G})$ on GL, defined by the group multiplication $\mu(\mathbf{F}, \mathbf{G}) := \mathbf{F}\mathbf{G}$, now induces on GL/O an operation of right translation $\hat{\rho}$ via

$$\hat{\rho}: GL \times GL/O \to GL/O \qquad (\mathbf{G}, [\mathbf{F}]) \mapsto [\rho_{\mathbf{G}}(\mathbf{F})] = [\mathbf{F}\mathbf{G}], \tag{21}$$

so that both parts of the diagram are commutative.

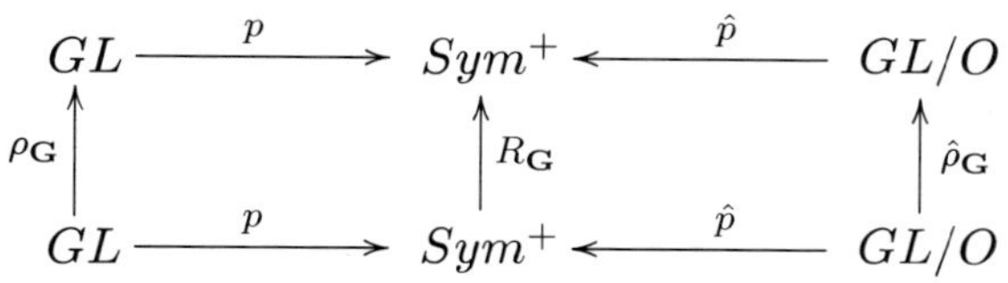

$$\begin{array}{ccccc}
GL & \xrightarrow{\;p\;} & Sym^+ & \xleftarrow{\;\hat{p}\;} & GL/O \\
\uparrow{\scriptstyle \rho_{\mathbf{G}}} & & \uparrow{\scriptstyle R_{\mathbf{G}}} & & \uparrow{\scriptstyle \hat{\rho}_{\mathbf{G}}} \\
GL & \xrightarrow{\;p\;} & Sym^+ & \xleftarrow{\;\hat{p}\;} & GL/O
\end{array}$$

That is

$$p \circ \rho_{\mathbf{G}} = R_{\mathbf{G}} \circ p \qquad \text{and} \qquad R_{\mathbf{G}} \circ \hat{p} = \hat{p} \circ \hat{\rho}_{\mathbf{G}}, \tag{22}$$

where $\hat{\rho}_{\mathbf{G}}(.) := \hat{\rho}(\mathbf{G}, .)$ and $R_{\mathbf{G}}(.) := R(\mathbf{G}, .)$ for any $\mathbf{G} \in GL$. This property, called *equivariance*, enables us instead of studying the action of GL on Sym^+ equally well to study the action of GL on GL/O, or GL on itself. This property then plays a principal role in numerical integration as discussed in Sect. 4

The configuration space Sym^+, like any other homogeneous spaces, possesses high degree of symmetries that preserve Riemannian metric, which can be naturally introduced on Sym^+ [13]: homogeneity, isotropy, and point reflection symmetry. Every point in such spaces looks like every other point, and so we speak of the *Riemannian (globally) symmetric spaces* [1, 2, 4, 20, 28]. These spaces are *constantly curved manifolds*, and represent a direct generalization of the Euclidean space. For more about Sym^+ from the viewpoint of Riemannian (globally) symmetric spaces, see especially chap. 5.4 in [20], and from the viewpoint of spaces of non-positive curvature (actually negative for Sym^+), chap. II.10 in [4].

2.3. Lie Algebras $\mathfrak{gl}(N)$ and $\mathfrak{o}(N)$ and Lie Triple System $sym(N)$

Though $Sym^+(N)$ is not a group, but the Riemannian symmetric space, it is still very closely linked with the Lie group theory via the right coset space (19) of the group $GL(N)$ modulo its closed subgroup $O(N)$. In addition, the Lie algebra $\mathfrak{gl}(N)$ of $GL(N)$ is a direct sum of two vector spaces: the Lie algebra $\mathfrak{o}(N)$ of all skew-symmetric matrices corresponding to orthogonal group $O(N)$, and the vector space of all symmetric matrices $sym(N)$

$$\mathfrak{gl}(N, \mathbb{R}) = \mathfrak{o}(N, \mathbb{R}) \oplus sym(N, \mathbb{R}). \tag{23}$$

If we introduce Lie bracket operation [19, 25]

$$[\mathbf{A}, \mathbf{B}] := \mathbf{AB} - \mathbf{BA}, \tag{24}$$

then $[\mathfrak{o}, \mathfrak{o}] \subset \mathfrak{o}$, $[sym, sym] \subset \mathfrak{o}$, $[\mathfrak{o}, sym] \subset sym$. This property means that the subspace sym forms in $\mathfrak{gl}$ the so-called *Lie triple system* with respect to the double commutator $[.\,,[.\,,.]]$ – for more see [1, 22, 28]. Similarly to (19), we have

$$T_{\mathbf{I}}Sym^+(N, \mathbb{R}) \simeq sym(N, \mathbb{R}) \simeq T_{[\mathbf{I}]}\, GL(N, \mathbb{R})/O(N, \mathbb{R}). \tag{25}$$

In addition, for Sym^+, contrary to $\exp : \mathfrak{gl} \to GL$, the matrix exponential

$$\exp : sym \to Sym^+ \qquad \mathbf{A} \mapsto \exp(\mathbf{A}) = \sum_{n \geq 0} \frac{1}{n!}\mathbf{A}^n \tag{26}$$

is moreover a global diffeomorphism together with its inverse.

3. Evolution Equation of Lie-type for Finite Deformations

Now, we are ready to analyse relation (12) on Sym^+

$$\partial \mathbf{C}_t = 2\mathbf{F}^T \mathbf{d} \mathbf{F}, \tag{27}$$

together with (14) on GL

$$\partial \mathbf{F}_t = \mathbf{L}\mathbf{F}_t, \tag{28}$$

where $2\mathbf{d} = \mathbf{L} + \mathbf{L}^T$, and $\mathbf{L} = \nabla v$ stands for the velocity gradient.

First, the rate-of-deformation tensor $\mathbf{d}$ is a mixed second-order tensor in actual configuration, and so its image in referential configuration

$$\mathbf{D} := \Phi^*(\mathbf{d}) \equiv \mathbf{F}^{-1}\mathbf{d}\mathbf{F}, \tag{29}$$

the *referential rate-of-deformation* tensor, is the pull-back of $\mathbf{d}$ induced by deformation Φ [13, 24]. Since $\mathbf{F}^T\mathbf{d}\mathbf{F} = \mathbf{C}\mathbf{F}^{-1}\mathbf{d}\mathbf{F} = \mathbf{C}\mathbf{D}$, we conclude that

$$\partial\mathbf{C}_t = 2\mathbf{C}_t\mathbf{D}\left(= 2\mathbf{D}^T\mathbf{C}_t\right) \in T_{\mathbf{C}_t}Sym^+, \tag{30}$$

where the second equality is due to symmetry.

On the other hand, if we employ the polar decomposition of the deformation gradient $\mathbf{F} = \mathbf{R}\mathbf{U}$ where $\mathbf{C} = \mathbf{U}^2$, we can introduce

$$\hat{\mathbf{D}} := \mathbf{R}^{-1}\mathbf{d}\mathbf{R}, \tag{31}$$

the *co-rotational (rotated) rate-of-deformation* tensor [13], so that relation (27) takes another form

$$\partial\mathbf{C}_t = 2\mathbf{U}_t\hat{\mathbf{D}}\mathbf{U}_t \in T_{\mathbf{C}_t}Sym^+. \tag{32}$$

Note, that if $\mathbf{D}$ resp. $\hat{\mathbf{D}}$ is a matrix constant, solution to (30) resp. (32) is a geodesic in Sym^+ [11, 13]. Both referential and co-rotational rate-of-deformation tensors play an important role in numerical incremental analysis of non-linear materials [26, 27].

3.1. Geometrical Interpretation

Let us now discuss the right-hand sides of the *evolution equations*

$$\partial_t\mathbf{C}_t = \mathbf{D}^T\mathbf{C} + \mathbf{C}\mathbf{D}, \qquad \text{where} \quad \mathbf{D} := \mathbf{F}^{-1}\mathbf{d}\mathbf{F} \tag{33}$$

$$= 2\mathbf{U}\hat{\mathbf{D}}\mathbf{U}, \qquad \text{where} \quad \hat{\mathbf{D}} := \mathbf{R}^{-1}\mathbf{d}\mathbf{R} \tag{34}$$

as well as

$$\partial_t\mathbf{F} = \mathbf{L}\mathbf{F}, \quad \text{where} \quad \mathbf{L} = \nabla v \quad \text{and} \quad 2\mathbf{d} = \mathbf{L} + \mathbf{L}^{\mathrm{T}} \tag{35}$$

from the viewpoint of geometrical properties of the underlying configuration space Sym^+, in particular those related to its Lie group of transformations GL.

Remind that at the infinitesimal level, the action of the Lie group induces an action of its Lie algebra expressed in terms of the fundamental vector fields

(4). We shall demonstrate that the right-hand sides of the relations (33) and (34) actually represent such fundamental vector fields indexed by the corresponding rate-of-deformation tensors $\mathbf{D}$ or $\hat{\mathbf{D}}$ from the Lie algebra $\mathfrak{gl}$ corresponding to the Lie group of transformations GL. We prove that the generating operation Ψ from (4) will be induced by the right translation R on Sym^+, see (17).

Let us denote by $R^{\mathbf{C}} := R(.,\mathbf{C})$, i.e.

$$R^{\mathbf{C}} : GL \to Sym^+ \qquad \mathbf{G} \mapsto \mathbf{G}^T\mathbf{C}\mathbf{G}, \tag{36}$$

and by $Q^{\mathbf{C}} := R(\mathbf{U},.)$, where $\mathbf{U}$ is restricted to $Sym^+ \subset GL$ and $\mathbf{C} = \mathbf{U}^2$,

$$Q^{\mathbf{C}} : Sym^+ \to Sym^+ \qquad \mathbf{B} \mapsto \mathbf{U}\mathbf{B}\mathbf{U}, \tag{37}$$

and recall $\rho_{\mathbf{F}}(.) := \mu(\,.\,,\mathbf{F})$ on GL induced by the group multiplication $\mu(\mathbf{G},\mathbf{F}) := \mathbf{G}\mathbf{F}$

$$\rho_{\mathbf{F}} : GL \to GL \qquad \mathbf{G} \mapsto \mathbf{G}\mathbf{F}, \tag{38}$$

then evidently

$$R^{\mathbf{C}}(\mathbf{I}) \equiv R(\mathbf{I},\mathbf{C}) = \mathbf{I}\mathbf{C}\mathbf{I} = \mathbf{C}\,, \tag{39}$$

$$Q^{\mathbf{C}}(\mathbf{I}) \equiv R(\mathbf{U},\mathbf{I}) = \mathbf{U}\mathbf{I}\mathbf{U} = \mathbf{C}\,, \tag{40}$$

$$\rho_{\mathbf{F}}(\mathbf{I}) \equiv \mu(\mathbf{I},\mathbf{F}) = \mathbf{I}\mathbf{F} = \mathbf{F}. \tag{41}$$

The corresponding tangent spaces at $\mathbf{I}$, for both GL and Sym^+, are mapped into $T_{\mathbf{C}}Sym^+$, resp. $T_{\mathbf{F}}GL$ by corresponding tangent map $T_{\mathbf{I}}R^{\mathbf{C}}$ and $T_{\mathbf{I}}Q^{\mathbf{C}}$, resp. $T\rho_{\mathbf{F}}$ via (4):

$$r_* \equiv T_{\mathbf{I}}R^{\mathbf{C}} : T_{\mathbf{I}}GL \to T_{\mathbf{C}}Sym^+ \qquad \mathbf{H} \mapsto \xi_{\mathbf{H}}(\mathbf{C}) = \mathbf{H}^T\mathbf{C} + \mathbf{C}\mathbf{H}, \tag{42}$$

$$q_* \equiv T_{\mathbf{I}}Q^{\mathbf{C}} : T_{\mathbf{I}}Sym^+ \to T_{\mathbf{C}}Sym^+ \qquad \hat{\mathbf{H}} \mapsto \xi_{\hat{\mathbf{H}}}(\mathbf{C}) = \mathbf{U}\hat{\mathbf{H}}\mathbf{U}, \tag{43}$$

$$\rho_* \equiv T_{\mathbf{I}}\rho_{\mathbf{F}} : T_{\mathbf{I}}GL \to T_{\mathbf{F}}GL \qquad \mathbf{J} \mapsto \xi_{\mathbf{J}}(\mathbf{F}) = \mathbf{J}\mathbf{F}. \tag{44}$$

which are applied on vectors

$$\mathbf{H} \in T_{\mathbf{I}}GL \equiv \mathfrak{gl} \qquad \text{where } \mathbf{H} = \partial\mathbf{G}_{t=0}, \mathbf{G}_{t=0} = \mathbf{I}, \tag{45}$$

$$\hat{\mathbf{H}} \in T_{\mathbf{I}}Sym^+ \equiv sym \subset \mathfrak{gl} \qquad \text{where } \hat{\mathbf{H}} = \partial\mathbf{B}_{t=0}, \mathbf{B}_{t=0} = \mathbf{I}\,, \tag{46}$$

$$\mathbf{J} \in T_{\mathbf{I}}GL \equiv \mathfrak{gl} \qquad \text{where } \mathbf{J} = \partial\mathbf{E}_{t=0}, \mathbf{E}_{t=0} = \mathbf{I}\,, \tag{47}$$

being defined in terms of curves $\mathbf{G}_t \subset GL$, $\mathbf{B}_t \subset Sym^+$ and $\mathbf{E}_t \subset GL$.

From (42) and (43), we thus conclude that $\partial\mathbf{C}_t$ can be expressed as the fundamental vector field indexed by time-dependent elements $\mathbf{D}_t \in \mathfrak{gl}$ or $\hat{\mathbf{D}}_t \in sym$ given by

$$\partial\mathbf{C}_t = \mathbf{D}_t^T\mathbf{C}_t + \mathbf{C}_t\mathbf{D}_t = r_*(\mathbf{D}_t)(\mathbf{C}_t) \qquad \mathbf{D}: \mathbb{R}^+ \times Sym^+ \to \mathfrak{gl} \tag{48}$$

$$= 2\mathbf{U}_t\hat{\mathbf{D}}_t\mathbf{U}_t \qquad = q_*(2\hat{\mathbf{D}}_t)(\mathbf{C}_t) \qquad \hat{\mathbf{D}}: \mathbb{R}^+ \times Sym^+ \to sym\,, \tag{49}$$

The time-indexed quantity $\mathbf{D}_t$ forms a curve in the Lie algebra of all matrices $\mathfrak{gl}$, while $\hat{\mathbf{D}}_t$ forms a curve in its linear subspace of all symmetric matrices sym. Similarly, due to (44), the relation for deformation gradient can be expressed as the fundamental vector field indexed by time-dependent elements $\mathbf{L}_t \in \mathfrak{gl}$

$$\partial\mathbf{F}_t = \mathbf{L}_t\mathbf{F}_t = \rho_*(\mathbf{L}_t)(\mathbf{F}_t) \qquad \mathbf{L}: \mathbb{R}^+ \times GL \to \mathfrak{gl}\,, \tag{50}$$

That is, the velocity gradients $\mathbf{L}_t$ form a curve in the Lie algebra $\mathfrak{gl}$, too.

Equations (48), (49) and (50) in the form of a fundamental vector field are known as the evolution **equations of Lie-type** [5, 17]. Since fundamental vector fields are transformed vectors from the Lie algebra corresponding to the Lie group of transformations on the configuration space, one can employ the whole Lie-group theory that offers for these equations efficient and geometrically consistent numerical methods for time-discrete integration.

3.2. Solving Strategy

When quantities $\mathbf{D}_t$, $\hat{\mathbf{D}}_t$ or $\mathbf{L}_t$ depend only on time t, evolution Eqs. (48), (49) and (50) are called *linear* ordinary differential equations, in contrast to the *non-linear* case when they additionally depend on the respective points $\mathbf{C}_t$ or $\mathbf{F}_t$ of the configuration space.

The time-dependent linear Eq. (48) is straightforward to reduce to an equation that is similar to (50).

Proposition 1. *If* $\mathbf{D}: \mathbb{R}^+ \to \mathfrak{gl}$ *is a curve in the Lie algebra of the Lie group* GL*, and* $\mathbf{G}: \mathbb{R}^+ \to GL$ *is the solution to the linear equation*

$$\partial\mathbf{G}_t = \mathbf{G}_t\mathbf{D}_t = \lambda_*(\mathbf{D}_t)(\mathbf{G}_t) \tag{51}$$

with initial condition $\mathbf{G}_0 = \mathbf{I}$*, then*

$$\mathbf{C}_t = R(\mathbf{G}_t, \mathbf{C}_0) \equiv \mathbf{G}_t^T\mathbf{C}_0\,\mathbf{G}_t \tag{52}$$

is the solution to the Eq. (50) with initial condition $\mathbf{C}_0$.

$\mathbf{G}_t$ is called a *fundamental solution*. Equation (51) is nothing but a modification of (50) by using the left translation on GL instead of its right counterpart. In fact, if we replace right translations $\rho_{\mathbf{F}}$ for the left ones $\lambda_{\mathbf{G}}(.) := \mu(\mathbf{G}, .)$, then cf. (44),

$$\lambda_* \equiv T_{\mathbf{I}}\lambda_{\mathbf{G}} : T_{\mathbf{I}}GL \to T_{\mathbf{G}}GL \qquad \mathbf{D} \mapsto \xi_{\mathbf{D}}(\mathbf{G}) = \mathbf{GD}. \tag{53}$$

Moreover, if $\mathbf{D} = \mathbf{D}_0$ is a constant matrix, then the solution to Eq. (51) is the *one-parameter subgroup* $\mathbf{G}_t = \exp t\mathbf{D}_0$ of GL, and the solution to Eq. (48) reads as

$$\mathbf{C}_t = R(\exp t\mathbf{D}_0, \mathbf{C}_0) \equiv (\exp t\mathbf{D}_0)^T \mathbf{C}_0 (\exp t\mathbf{D}_0). \tag{54}$$

Similarly, the *one-parameter subgroup* $\mathbf{F}_t = \exp t\mathbf{L}_0$ also solves Eq. (50), provided $\mathbf{L} = \mathbf{L}_0$ is a constant matrix. The solution passes through $\mathbf{I}$, and to ensure the initial condition $\mathbf{F}_0$, it has to be translated by means of $\rho_{\mathbf{F}_0}$

$$\mathbf{F}_t = \rho_{\mathbf{F}_0}(\exp t\mathbf{L}_0) \equiv (\exp t\mathbf{L}_0)\mathbf{F}_0. \tag{55}$$

Time-dependent (non-autonomous) linear Eqs. (50) and (51) are usually solved by Magnus or Fer expansion, see Sect. 6.

Equation (49) is a slightly different case in that there is a global coordinate system established via the matrix exponential exp (26)– a one-to-one differentiable mapping between the vector space sym and the configuration space Sym^+. As a result, sym can serve as an auxiliary vector space for intermediate numerical time-discrete integration, see Sect. 4.2. Moreover, if $\hat{\mathbf{D}} = \hat{\mathbf{D}}_0$ is a constant matrix, then solution to (49)

$$\mathbf{C}(t) = \mathbf{U}_0 \exp(2\, t\hat{\mathbf{D}}_0)\mathbf{U}_0 = \mathbf{C}_0 \exp(t\, \mathbf{C}_0^{-1}\partial\mathbf{C}_0) \tag{56}$$

is a *geodesic* in Sym^+ [11, 13].

In solid mechanics, however, we encounter general *nonlinear equations*, and we have to resort to numerical solution. The objective is to seek an approximation $\mathbf{C}_{n+1} \approx \mathbf{C}(t_{n+1}) \in Sym^+$ from a given known approximation $\mathbf{C}_n \approx \mathbf{C}(t_n) \in Sym^+$. Since Sym^+ is a homogeneous manifold with respect to the transitive action of group GL, we can employ its underlying geometry as suggested by (52) in Proposition 1, and search for an element $\mathbf{G}_n(h; \mathbf{C}_n, ...) \in GL$, close to the identity matrix $\mathbf{I}$, such that the update reads as

$$\mathbf{C}_{n+1} = R(\mathbf{G}_n(h), \mathbf{C}_n) \in Sym^+. \tag{57}$$

This procedure then automatically guarantees that the numerical solution stays in Sym^+ at any point of time. With Eqs. (49) and (50) we proceed in the similar way, making use of (55) and (56).

In the case of Sym^+, contrary to the classical numerical integration of ordinary differential equations evolving on $\mathbb{R}^N$

$$\boldsymbol{y}' = \boldsymbol{f}(t, \boldsymbol{y}), \qquad t \geq 0, \qquad \boldsymbol{y}(0) = \boldsymbol{y}_0, \qquad \boldsymbol{y}(t) \in \mathbb{R}^N, \tag{58}$$

where $\boldsymbol{f}$ is a vector field on $\mathbb{R}^+ \times \mathbb{R}^N$, we can no longer rest upon the linear structure of the vector space $\mathbb{R}^N$. The thing is that such a structure makes it possible to identify the configuration space $\mathbb{R}^N$ with its tangent spaces $T_{\boldsymbol{y}}\mathbb{R}^N \simeq \mathbb{R}^N$, and notably the group of motions–given by *translations* δ on $\mathbb{R}^N$, with $\mathbb{R}^N$ itself. This all facilitates the numerical solution. In fact, classical numerical integrators enable to advance a time-stepping procedure by adding vectors $\boldsymbol{a}$ to points $\boldsymbol{y}$ in $\mathbb{R}^N$

$$\boldsymbol{y}_{n+1} = \boldsymbol{y}_n + h\boldsymbol{a}_n = \delta(\boldsymbol{a}_n(h), \boldsymbol{y}_n) \equiv \delta^{\boldsymbol{y}_n}(\boldsymbol{a}_n(h)), \tag{59}$$

where $\boldsymbol{a}_n = \boldsymbol{a}_n(h\,; \boldsymbol{y}_n, ...)$ can be computed, for example, by the Runge–Kutta numerical method.

A detailed comparison of (57) with (59) suggests that if we properly distinguish the configuration space $Sym^+ = GL/O$ resp. GL from its group of motions GL, as well as from their corresponding tangent spaces $T_{\mathbf{C}}Sym^+ \simeq sym$ resp. $T_{\mathbf{F}}GL \simeq \mathfrak{gl}$, compared to $\mathbb{R}^N$ in (59), we can still modify the classical numerical integrators to our purposes. We shall make it clear in the next section, wherein we derive associated evolution equations on the auxiliary vector spaces $\mathfrak{gl}$ and sym with the underlying linear structure.

4. Equation of Lie-Type on the Lie Algebra

A particular class of geometric numerical integrators has been formulated for Lie groups and homogeneous spaces [7, 17, 21], wherein the classical procedures have been revised due to their relations to Lie algebras with the additional algebraic structure of the Lie bracket (24).

One of the great virtues of these methods derives from the fact that the local structure in a neighborhood of any point of a Lie group G can be described by its Lie algebra $\mathfrak{g}$. In fact, the tangent space $T_gG \simeq T_eG \equiv \mathfrak{g}$ for all $g \in G$ due to left/right translations, resulting in the left/right trivialized tangent bundle

$TG \simeq G \times \mathfrak{g}$, see (70)-(71). Since the Lie algebras are closed not only under linear combinations, which is crucial for the application of classical numerical methods, but also under the Lie bracket operation (24), working in Lie algebras enables to employ some very technical tools from Lie-algebra theory that can be used to great effect in reducing the cost of numerical approximations.

A transition between the Lie algebra and the Lie group is then accomplished by a differentiable mapping $\Psi : \mathfrak{g} \to G$, which maps the zero element $0 \in \mathfrak{g}$ to the identity $e \in G$, thus establishing a coordinate system in the group by left/right translations. Usually, Ψ stands for the exponential mapping (26).

A common procedure – referred to in Proposition 1, that preserves the associated structure of the configuration space via relation (57), is to start with the fundamental solution $g(t)$ and transform it from the Lie group G to the underlying Lie algebra $\mathfrak{g}$ by an inverse of the differentiable mapping Ψ. Then, we can formulate and solve an associated differential equation in $\mathfrak{g}$, and finally map the solution back to G. In this way, the discretization procedure is performed in the Lie algebra, and so the classical Runge – Kutta method for approximation of the solution to the associated differential equation on $\mathfrak{g}$ can be employed. This approach, as suggested by Munthe-Kaas [21], results in class of schemes called the Runge – Kutta – Munthe-Kaas methods.

We shall employ common features of the evolution equations of Lie-type (48), (49) and (50), first by transforming them locally into differential equations on the Lie algebra $\mathfrak{gl}$, which corresponds to the Lie group GL acting on the homogeneous space Sym^+. For general discussion, consult [9].

4.1. Equation $\partial \mathbf{C}_t = \mathbf{D}_t^T \mathbf{C}_t + \mathbf{C}_t \mathbf{D}_t$

Let us start by analysing the equation

$$\partial \mathbf{C}_t = \mathbf{D}_t^T \mathbf{C}_t + \mathbf{C}_t \mathbf{D}_t \equiv r_*(\mathbf{D}_t)(\mathbf{C}_t) \qquad \mathbf{D} : \mathbb{R}^+ \times Sym^+ \to \mathfrak{gl} \tag{60}$$

with the initial condition $\mathbf{C}_0$. Since we shall be considering a nonlinear type of the equation, in which $\mathbf{D}_t$ depends also on $\mathbf{C}_t$, we have to proceed incrementally.

As discussed in Sect. 3.2, the task to find an update $\mathbf{C}_{n+1}$ for each time increment $t_{n+1} = t_n + h$ reduces in the neighborhood of the previous approximation $\mathbf{C}_n \approx \mathbf{C}(t_n)$ to solving the Eq. (51) on GL

$$\partial \mathbf{G}_t = \mathbf{G}_t \mathbf{D}(t, \mathbf{C}_t) \tag{61}$$

by an element $\mathbf{G}_n(h; \mathbf{C}_n, ...) \in GL$ close to the identity matrix $\mathbf{I}$, such that the new update (57) reads

$$\mathbf{C}_{n+1} = R(\mathbf{G}_n(h), \mathbf{C}_n) \in Sym^+. \tag{62}$$

Since the exponential mapping exp is local diffeomorphism between $\mathfrak{gl}$ and GL, we can, instead of $\mathbf{G}_t$, search for a curve $\mathbf{\Theta}_t = \log \mathbf{G}_t$ in the Lie algebra $\mathfrak{gl}$, which is contained in a neighborhood of its origin $\mathbf{\Theta}_0 \equiv \mathbf{0} = \log \mathbf{I} \in \mathfrak{gl}$.

To formulate the corresponding evolution equation for $\mathbf{\Theta}_t$ in $\mathfrak{gl}$, it suffices to work out how to transform vectors $\partial \mathbf{C}_t$ on Sym^+ to vectors on $\mathfrak{gl}$. For that purpose we exploit the fact that the deformation rates (60) are expressed as the fundamental vector fields (4).

As a first step, we find an action b of GL on $\mathfrak{gl}$ such that the matrix exponential will be an equivariant map with respect to this action on $\mathfrak{gl}$ and the right translation ρ of GL on itself. Then, the action b will also correspond to the right translation R of GL on Sym^+ as depicted in diagram bellow, cf. diagram (2.2) in [9].

$$\begin{array}{ccccc} \mathfrak{gl} & \xrightarrow{\exp} & GL & \xrightarrow{R^{\mathbf{C}_n}} & Sym^+ \\ b_{\mathbf{g}} \uparrow & & \uparrow \rho_{\mathbf{g}} & & \uparrow R_{\mathbf{g}} \\ \mathfrak{gl} & \xrightarrow[\exp]{} & GL & \xrightarrow[R^{\mathbf{C}_n}]{} & Sym^+ \end{array}$$

A requirement of equivariance means that both parts of the diagram are commutative, i.e.

$$\exp \circ\, b_{\mathbf{g}}(\mathbf{\Theta}) = \rho_{\mathbf{g}} \circ \exp(\mathbf{\Theta}) \tag{63}$$

$$R^{\mathbf{C}_n} \circ \rho_{\mathbf{g}}(\mathbf{G}) = R_{\mathbf{g}} \circ R^{\mathbf{C}_n}(\mathbf{G}), \tag{64}$$

and so (63) defines the action b. Then, we turn to the infinitesimal description of the group actions. This enables us to introduce the corresponding fundamental vector field on $\mathfrak{gl}$ that will represent the transformed vector $\partial \mathbf{C}_t$.

Let $\gamma(s)$ be a curve in GL such that $\gamma(0) = \mathbf{I}$ and $\partial\gamma(0) = \mathbf{D} \in \mathfrak{gl}$. Note that the fundamental vector field (60) on Sym^+, cf. (42), can be expressed in terms of $R_{\mathbf{g}}$

$$\xi_{\mathbf{D}}(\mathbf{C}) := \frac{\partial}{\partial s}\bigg|_{s=0} R^{\mathbf{C}}(\gamma(s)) = \frac{\partial}{\partial s}\bigg|_{s=0} R_{\gamma(s)}(\mathbf{C}) \in T_{\mathbf{C}} Sym^+, \tag{65}$$

which enables to introduce the corresponding fundamental vector field $\delta_{\mathbf{D}}$ on GL in terms of $\rho_{\mathbf{g}}$

$$\delta_{\mathbf{D}}(\mathbf{G}) := \left.\frac{\partial}{\partial s}\right|_{s=0} \rho_{\gamma(s)}(\mathbf{G}) = \left.\frac{\partial}{\partial s}\right|_{s=0} \mathbf{G}\gamma(s) = \mathbf{GD} \in T_{\mathbf{G}}GL\,. \tag{66}$$

The searched fundamental vector field on $\mathfrak{gl}$ for $\mathbf{\Theta}$ close to the origin

$$\beta_{\mathbf{D}}(\mathbf{\Theta}) := \left.\frac{\partial}{\partial s}\right|_{s=0} b_{\gamma(s)}(\mathbf{\Theta}) \in T_{\mathbf{\Theta}}\mathfrak{gl} \simeq \mathfrak{gl} \tag{67}$$

is thus similarly expressed in terms of the operation $b_{\mathbf{g}}$, cf. diagram.

Due to Theorem 3.3 in [9], for $\mathbf{G} = \exp(\mathbf{\Theta})$, we get from (63) its local equivalent

$$T_{\mathbf{\Theta}} \exp \circ \beta_{\mathbf{D}}(\mathbf{\Theta}) = \delta_{\mathbf{D}} \circ \exp(\mathbf{\Theta}), \tag{68}$$

and similarly from (64)

$$T_{\mathbf{G}} R^{\mathbf{C}_n} \circ \delta_{\mathbf{D}}(\mathbf{G}) = \xi_{\mathbf{D}} \circ R^{\mathbf{C}_n}(\mathbf{G}), \tag{69}$$

where $T\exp : T\mathfrak{gl} \to TGL$, resp. $TR^{\mathbf{C}_n} : TGL \to TSym^+$ denote maps between corresponding tangent bundles at corresponding points $\mathbf{\Theta}$, resp. $\mathbf{G}$.

In other words, the following diagram for the fundamental vector fields $\beta_{\mathbf{D}}$, $\delta_{\mathbf{D}}$ and $\xi_{\mathbf{D}}$

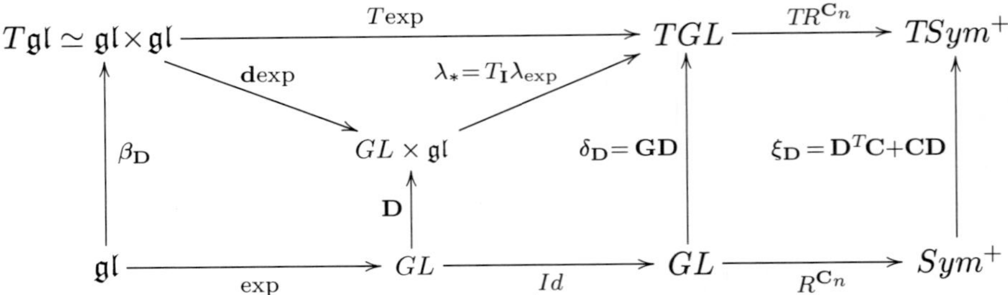

is commutative. We say that $\beta_{\mathbf{D}}$ and $\delta_{\mathbf{D}}$ are exp-related, and vector fields $\delta_{\mathbf{D}}$ and $\xi_{\mathbf{D}}$ are $R^{\mathbf{C}_n}$-related. Since fundamental vector fields are infinitesimal versions of Lie-group transformations (infinitesimal generators), the diagram represents an infinitesimal version of the previous one.

Let us now comment on the inner part of the diagram, in which Id stands for the identity map on GL. Since $\mathfrak{gl}$ is vector space, $T\mathfrak{gl} \simeq \mathfrak{gl} \times \mathfrak{gl}$, just like

$T\mathbb{R}^N \simeq \mathbb{R}^N \times \mathbb{R}^N$. On the other side, since a neighborhood of any point of Lie group looks locally the same, a tangent space $T_{\mathbf{G}}GL$ can be identified with $\mathfrak{gl}$ through isomorphism – here by the tangent map $T_{\mathbf{I}}\lambda_{\mathbf{G}} : \mathfrak{gl} \to T_{\mathbf{G}}GL$ induced by the left translation $\lambda_{\mathbf{G}}$, cf. (53). That is,

$$GL \times \mathfrak{gl} \equiv GL \times T_{\mathbf{I}}GL \approx TGL \tag{70}$$

$$(\mathbf{G}, \mathbf{D}) \mapsto T_{\mathbf{I}}\lambda_{\mathbf{G}}(\mathbf{D}) = \mathbf{GD} \equiv \delta_{\mathbf{D}}(\mathbf{G}), \tag{71}$$

which is called the left trivialization of the tangent bundle TGL. Compare also (71) with (66). Left translations $\lambda_{\mathbf{G}}$ come from relation, cf. (62),

$$R^{\mathbf{C}}(\mathbf{g}) = R^{\mathbf{C}_n} \circ \lambda_{\mathbf{G}}(\mathbf{g}), \tag{72}$$

$$\text{where } \mathbf{C} = R(\mathbf{G}, \mathbf{C}_n) \equiv (\mathbf{G})^T \mathbf{C}_n(\mathbf{G}), \tag{73}$$

when transforming the element $\mathbf{G}$ together with its close neighborhood made up of g, in order to find transformation of the corresponding fundamental vector fields. Then

$$T_{\mathbf{I}}R^{\mathbf{C}} = T_{\mathbf{G}}R^{\mathbf{C}_n} \circ T_{\mathbf{I}}\lambda_{\mathbf{G}} \tag{74}$$

$$\mathbf{D} \mapsto T_{\mathbf{G}}R^{\mathbf{C}_n}\big(\delta_{\mathbf{D}}(\mathbf{G})\big) = \xi_{\mathbf{D}}(\mathbf{C})\,. \tag{75}$$

Setting the left-trivialized tangent map $\mathbf{dexp}_{\mathbf{\Theta}} : \mathfrak{gl} \to \mathfrak{gl}$ of the matrix exponential exp via a decomposition of its tangent map at a point $\mathbf{\Theta} \in \mathfrak{gl}$ close to the origin:

$$T_{\mathbf{\Theta}} \exp = T_{\mathbf{I}}\lambda_{\exp \mathbf{\Theta}} \circ \mathbf{dexp}_{\mathbf{\Theta}}, \tag{76}$$

we get for the left-hand side of relation (68)

$$T_{\mathbf{\Theta}} \exp \circ \beta_{\mathbf{D}} = T_{\mathbf{I}}\lambda_{\exp \mathbf{\Theta}} \circ \mathbf{dexp}_{\mathbf{\Theta}} \circ \beta_{\mathbf{D}} \tag{77}$$

whereas for the right-hand side we have

$$\delta_{\mathbf{D}} \circ \exp \mathbf{\Theta} = T_{\mathbf{I}}\lambda_{\exp \mathbf{\Theta}} \circ \mathbf{D} \tag{78}$$

due to commutativity of the diagram above. Thus, the transformation formula for vector field $\beta_{\mathbf{D}}$ on $\mathfrak{gl}$ reads

$$\mathbf{dexp}_{\mathbf{\Theta}} \circ \beta_{\mathbf{D}} = \mathbf{D}. \tag{79}$$

Eventually, we conclude with the evolution equation for $\boldsymbol{\Theta}_t$ in the Lie algebra $\mathfrak{gl}$ with the initial condition $\boldsymbol{\Theta}_0 = \mathbf{0}$

$$\partial\boldsymbol{\Theta}_t = \beta_{\mathbf{D}_t}(\boldsymbol{\Theta}_t) \equiv \mathbf{dexp}^{-1}_{\boldsymbol{\Theta}_t}\big(\mathbf{D}(t, \mathbf{C}_t)\big) \qquad \mathbf{C}_t = R(\exp\boldsymbol{\Theta}_t, \mathbf{C}_n), \tag{80}$$

where

$$\mathbf{dexp}^{-1}_{\boldsymbol{\Theta}}(\mathbf{D}) := \mathbf{D} - \tfrac{1}{2}\,[\boldsymbol{\Theta}, \mathbf{D}] + \tfrac{1}{12}\,[\boldsymbol{\Theta}, [\boldsymbol{\Theta}, \mathbf{D}]] + ... \tag{81}$$

$$= \sum_{j=0}^{\infty} \frac{B_j}{j\,!}\,[\boldsymbol{\Theta}, [\boldsymbol{\Theta}, [... , [\boldsymbol{\Theta}, \mathbf{D}]...]]] \tag{82}$$

and B_j are j'th Bernoulli numbers. The first few coefficients are

$$\frac{B_j}{j\,!} = \begin{cases} 0 & \text{for } k \text{ odd, and } k \neq 1 \\ 1, -\frac{1}{2}, \frac{1}{12}, -\frac{1}{720}, \frac{1}{30240}, -\frac{1}{1209600} & \text{for } k = 0, 1, 2, 4, 6, 8, ... \end{cases} \tag{83}$$

see [9] and references therein. The $j \geq 1$ terms and the nonlinearity of (81) are the cost of non-commutativity of the Lie bracket $[\boldsymbol{\Theta}, \mathbf{D}] := \boldsymbol{\Theta}\mathbf{D} - \mathbf{D}\boldsymbol{\Theta}$ in $\mathfrak{gl}$.

By solving Eq. (80), we find a curve $\boldsymbol{\Theta}_t$ in the Lie algebra $\mathfrak{gl}$ through the origin $\mathbf{0}$, which locally solves the original evolution Eq. (27) or (60) in Sym^+ in terms of a curve

$$\mathbf{C}_t = R^{\mathbf{C}_n}(\exp\boldsymbol{\Theta}_t) = R(\exp\boldsymbol{\Theta}_t, \mathbf{C}_n) \equiv (\exp\boldsymbol{\Theta}_t)^T\mathbf{C}_n(\exp\boldsymbol{\Theta}_t) \tag{84}$$

through $\mathbf{C}_n$ for t close to t_n, cf. (62).

4.2. Equation $\partial\mathbf{C}_t = 2\mathbf{U}_t\hat{\mathbf{D}}_t\mathbf{U}_t$

The case of equation

$$\partial\mathbf{C}_t = 2\mathbf{U}_t\hat{\mathbf{D}}_t\mathbf{U}_t = q_*(2\hat{\mathbf{D}}_t)(\mathbf{C}_t) \qquad \hat{\mathbf{D}} : \mathbb{R}^+ \times Sym^+ \to sym \tag{85}$$

is slightly different in that it is not related to the group of transformations but to the global coordinate system established on Sym^+ by the exponential mapping $\exp : sym \to Sym^+$ (26). Contrary to the case of $\mathfrak{gl}$ and GL, the mapping is a global and bijective diffeomorphism. The situation is exactly the same as with the Cayley coordinates, which exist for several matrix Lie groups, in particular for the orthogonal group $O(N, \mathbb{R})$ describing rotation of rigid bodies

[8, 23]. The appealing feature of these coordinate systems is the lesser expense in computing $\mathrm{dexp}_{\boldsymbol{\Theta}}^{-1}$, which, moreover, can be determined exactly.

Now, by definition of tangent map $T\exp : Tsym \to TSym^{+}$, for respective vector fields β and ξ

$$T_{\boldsymbol{\Theta}} \exp \circ \beta(\boldsymbol{\Theta}) = \xi \circ \exp(\boldsymbol{\Theta}), \tag{86}$$

which means that the following diagram is commutative.

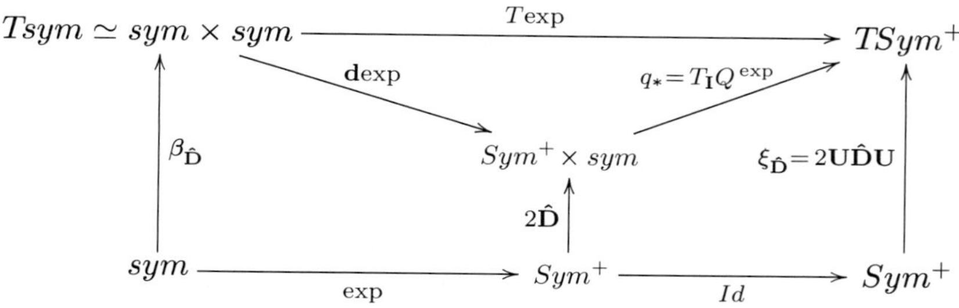

Since sym is a vector space, again $Tsym \simeq sym \times sym$.

As for the inner part of the diagram and indexation by $\hat{\mathbf{D}}$, the tangent space $T_{\mathbf{C}}Sym^{+}$ for any $\mathbf{C} \in Sym^{+}$ can be identified via (43) with sym, and so $TSym^{+}$ can be trivialised by

$$Sym^{+} \times sym \equiv Sym^{+} \times T_{\mathbf{I}}Sym^{+} \approx TSym^{+} \tag{87}$$

$$(\mathbf{C}, 2\hat{\mathbf{D}}) \longmapsto \quad T_{\mathbf{I}}Q^{\mathbf{C}}(2\hat{\mathbf{D}}) \quad = 2\mathbf{U}\hat{\mathbf{D}}\mathbf{U}. \tag{88}$$

Making use of the trivialized tangent $\mathrm{dexp}_{\boldsymbol{\Theta}}$ of the exponential mapping, cf. (76), via

$$T_{\boldsymbol{\Theta}} \exp = T_{\mathbf{I}}Q^{\exp \boldsymbol{\Theta}} \circ \mathrm{dexp}_{\boldsymbol{\Theta}}, \tag{89}$$

we obtain the transformed equation in sym

$$\partial \boldsymbol{\Theta}_t = \beta_{\hat{\mathbf{D}}_t}(\boldsymbol{\Theta}_t) \equiv \mathrm{dexp}_{\boldsymbol{\Theta}_t}^{-1}\big(2\hat{\mathbf{D}}(t, \mathbf{C}_t)\big) \qquad \mathbf{C}_t = \exp \boldsymbol{\Theta}_t, \tag{90}$$

with the initial condition $\boldsymbol{\Theta}_0 = \log \mathbf{C}_0$, in general different from $\mathbf{0}$.

In this case of $\exp : sym \to Sym^{+}$, the inverse of its trivialized tangent dexp^{-1} can be expressed exactly [2]. Namely, let $\boldsymbol{\Theta} = \mathbf{R}\boldsymbol{\Lambda}\mathbf{R}^{T} = \sum \lambda_i P_i$

be spectral decomposition, $\mathbf{\Lambda}$ the corresponding diagonal matrix with diagonal entries λ_i, and P_i corresponding projectors, then

$$\mathrm{dexp}_{\mathbf{\Theta}}^{-1}(\hat{\mathbf{D}}) := \mathbf{R}\left[\log^{[1]}(\mathbf{\Lambda}) \circ (\mathbf{R}^T\hat{\mathbf{D}}\mathbf{R})\right]\mathbf{R}^T \tag{91}$$

$$= \sum_i \sum_j \log^{[1]}(\lambda_i, \lambda_j) P_i \hat{\mathbf{D}} P_j , \tag{92}$$

where the *Hadamard (or Schur) product* $\mathbf{A} \circ \mathbf{B}$ of two matrices $\mathbf{A}$ and $\mathbf{B}$ is defined to be the matrix whose (i, j)-entry is $A_i^j B_i^j$, and where the 3×3 symmetric matrix $\log^{[1]}(\mathbf{\Lambda})$ has numbers

$$\log^{[1]}(\lambda_i, \lambda_j) = \frac{\log(\lambda_i) - \log(\lambda_j)}{\lambda_i - \lambda_j} \qquad \text{if } i \neq j \tag{93}$$

$$\log^{[1]}(\lambda_i, \lambda_i) = \log{}'(\lambda_i) = \frac{1}{\lambda_i} \tag{94}$$

as its (i, j)-entries. At point $\mathbf{0} \in sym$, the map $\mathrm{dexp}_{\mathbf{0}}^{-1} = \mathbf{I}$.

By solving Eq. (90), we find a curve $\mathbf{\Theta}_t$ on $sym \subset \mathfrak{gl}$ which solves the original evolution Eq. (27) or (85) in Sym^+ through the initial condition $\mathbf{C}_0$ in terms of a curve $\mathbf{C}_t = \exp \mathbf{\Theta}_t$ in Sym^+. Since $\mathrm{dexp}_{\mathbf{0}}^{-1} = \mathbf{I}$, the precision of the solution at $\mathbf{C}_0$ can be locally increased, provided we first transform an initial condition $\mathbf{C}_0$ to $\mathbf{I}$ by operation (37), then solve the Eq. (90) through $\mathbf{0}$, and move the local solution $\mathbf{C}_t = \exp \mathbf{\Theta}_t$ through $\mathbf{I}$ back to $\mathbf{C}_0$

$$\mathbf{C}_t = Q^{\mathbf{C}_0}(\exp \mathbf{\Theta}_t) \equiv \mathbf{C}_0^{1/2}(\exp \mathbf{\Theta}_t)\mathbf{C}_0^{1/2} \tag{95}$$

in each increment.

4.3. Equation $\partial \mathbf{F}_t = \mathbf{L}_t \mathbf{F}_t$

The evolution equation for deformation gradient

$$\partial \mathbf{F}_t = \mathbf{L}_t \mathbf{F}_t = \rho_*(\mathbf{L}_t)(\mathbf{F}_t) \qquad \mathbf{L} : \mathbb{R}^+ \times GL \to \mathfrak{gl} \tag{96}$$

is practically the same as the equation for the fundamental solution (51), except that the left translation is replaced by the right one. Even though $\mathbf{F}_t$ can now attain any value from GL, still we can proceed the same way provided we replace $\mathbf{F}_t$ by $\mathbf{G}_t = \mathbf{F}_t \mathbf{F}_n^{-1}$ and Eq. (96) by

$$\partial \mathbf{G}_t = \mathbf{L}_t \mathbf{G}_t = \rho_*(\mathbf{L}_t)(\mathbf{G}_t) \qquad \mathbf{L} : \mathbb{R}^+ \times GL \to \mathfrak{gl}. \tag{97}$$

Since the initial condition $\mathbf{F}(0) = \mathbf{F}_0$ now reads $\mathbf{G}(0) = \mathbf{I}$, we can directly apply approach from Sect. 4.1, captured in the diagram, cf. (70) and (71):

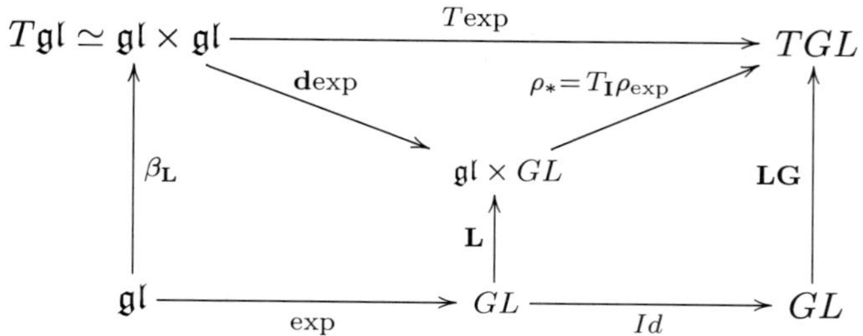

Making use of the right-trivialized tangent map $\mathbf{d}\exp_{\mathbf{\Theta}} : \mathfrak{gl} \to \mathfrak{gl}$, cf. (76), via a decomposition of the tangent map

$$T_{\mathbf{\Theta}} \exp = T_{\mathbf{I}}\rho_{\exp \mathbf{\Theta}} \circ \mathbf{d}\exp_{\mathbf{\Theta}}\,, \tag{98}$$

we again conclude that

$$\partial \mathbf{\Theta}_t = \beta_{\mathbf{L}_t}(\mathbf{\Theta}_t) \equiv \mathbf{d}\exp^{-1}_{\mathbf{\Theta}_t}\big(\mathbf{L}(t, \mathbf{G}_t)\big) \qquad \mathbf{G}_t = \exp \mathbf{\Theta}_t, \tag{99}$$

which results in a local solution to the Eq. (96) satisfying the initial condition $\mathbf{F}_0$

$$\mathbf{F}_t = \rho_{\mathbf{F}_0}(\exp \mathbf{\Theta}_t) \equiv (\exp \mathbf{\Theta}_t)\mathbf{F}_0. \tag{100}$$

The inverse of the trivialized tangent $\mathbf{d}\exp^{-1}$ is given by (81)-(83).

5. Runge – Kutta – Munthe-Kaas Method

The classical ν-stage Runge – Kutta method [14, 15] is defined by constants $\{c_k\}_{k=1}^{\nu}$, $\{a_{k,l}\}_{k,l=1}^{\nu}$, $\{b_l\}_{l=1}^{\nu}$, which are usually written as the *Butcher tableau*:

$$\begin{array}{c|c} c & A \\ \hline & b^T \end{array} \quad = \quad \begin{array}{c|cccc} c_1 & a_{1,1} & a_{1,2} & \cdots & a_{1,\nu} \\ c_2 & a_{2,1} & a_{2,2} & \cdots & a_{2,\nu} \\ \vdots & \vdots & \vdots & \ddots & \vdots \\ c_\nu & a_{\nu,1} & a_{\nu,2} & \cdots & a_{\nu,\nu} \\ \hline & b_1 & b_2 & \cdots & b_\nu \end{array} \tag{101}$$

When applied to the standard vector equation $\boldsymbol{y}' = \boldsymbol{f}(t, \boldsymbol{y})$ on $\mathbb{R}^N$ (58), a single step of length h from $\boldsymbol{y}_n = \boldsymbol{y}(t_n)$ to updated $\boldsymbol{y}_{n+1} = \boldsymbol{y}(t_n + h)$ is given by first solving following system of equations for $\boldsymbol{f}_l$

$$\left.\begin{aligned} \boldsymbol{f}_k = \boldsymbol{f}(t_n + c_k h, \boldsymbol{y}_n + h\boldsymbol{\theta}_k) \\ \text{where } \boldsymbol{\theta}_k = \textstyle\sum_{l=0}^{\nu} a_{k,l}\boldsymbol{f}_l \end{aligned}\right\} \quad k = 1, \ldots, \nu \tag{102}$$

and then followed by (59)

$$\boldsymbol{y}_{n+1} = \boldsymbol{y}_n + h\,\boldsymbol{\theta}, \quad \text{where } \boldsymbol{\theta} = \sum_{l=0}^{\nu} b_l \boldsymbol{f}_l \,. \tag{103}$$

Recall that $\boldsymbol{y}$ is a point whereas $\boldsymbol{f}(t, \boldsymbol{y})$ is a vector at $\boldsymbol{y}$, and that vectors can be translated to any point in $\mathbb{R}^N$ without a change. As we have already mentioned in Sect. 3.2, the vector space $\mathbb{R}^N$ plays a triple role here: the space of points, the additive group of translation operating on the space of point, and finally, the vector space identified with the original space $\mathbb{R}^N$ on which the actual integration is carried out.

For the configuration space Sym^+ with its group of translations GL, however, this is not the case, and we have to resort to an auxiliary linear vector space being naturally isomorphic to $\mathbb{R}^N$, provided we properly transform all the quantities and operations there and back. The previous section suggests two of them: either the Lie algebra of all matrices $\mathfrak{gl}$ or the vector space of all symmetric matrices $sym \subset \mathfrak{gl}$. Following [21], we point out the role of actions (36) and (37) for construction of a modified Runge – Kutta method on the homogeneous space Sym^+, which consists in searching for such an group element whose action induces desired approximation (57). We proceed similarly for the Lie group GL.

While analysing relations (27), resp. (28) in Sect. 3, we have restated them in a generic form of the evolution equations of Lie-type, cf. (48), (49), resp. (50),

$$\partial\boldsymbol{\Upsilon}_t = \phi_*(\boldsymbol{\Delta}_t)(\boldsymbol{\Upsilon}_t), \qquad t \geq 0, \qquad \boldsymbol{\Upsilon}(0) = \boldsymbol{\Upsilon}_0, \tag{104}$$

where $\boldsymbol{\Upsilon}$ stands for either $\mathbf{C}$ or $\mathbf{F}$, and respectively

$$\phi_* = r_* : \mathfrak{gl} \times Sym^+ \to TSym^+ \quad \text{with} \quad \boldsymbol{\Delta}_t = \mathbf{D}_t \subset \mathfrak{gl}, \tag{105}$$

$$\phi_* = q_* : sym \times Sym^+ \to TSym^+ \quad \text{with} \quad \boldsymbol{\Delta}_t = 2\hat{\mathbf{D}}_t \subset sym, \tag{106}$$

$$\phi_* = \rho_* : \mathfrak{gl} \times GL \to TGL \quad \text{with} \quad \boldsymbol{\Delta}_t = \mathbf{L}_t \subset \mathfrak{gl}. \tag{107}$$

Since Sym^+ and GL are homogeneous manifolds with respect to the transitive action of group GL, then instead of seeking an approximation on the homogeneous space directly, we can search for an element $\mathbf{G}_n(h) \in GL$, which depends on the time-step $h > 0$, such that the update reads, see (36), (37) and (38)

$$\mathbf{C}_{n+1} = R^{\mathbf{C}_n}(\mathbf{G}_n(h)) \equiv R(\mathbf{G}_n(h), \mathbf{C}_n) = \mathbf{G}_n(h)^T \mathbf{C}_n \, \mathbf{G}_n(h) \in Sym^+ \text{ or} \tag{108}$$

$$\mathbf{C}_{n+1} = Q^{\mathbf{C}_n}(\mathbf{G}_n(h)) \equiv R(\mathbf{C}_n^{1/2}, \mathbf{G}_n(h)) = \mathbf{C}_n^{1/2}\mathbf{G}_n(h)\mathbf{C}_n^{1/2} \in Sym^+, \tag{109}$$

$$\mathbf{F}_{n+1} = \rho_{\mathbf{F}_n}(\mathbf{G}_n(h)) \equiv \rho(\mathbf{G}_n(h), \mathbf{F}_n) = \mathbf{G}_n(h)\mathbf{F}_n \in GL\,. \tag{110}$$

This procedure then automatically guarantees that the numerical solution stays in Sym^+ or GL at any point of time.

Second, since $\mathbf{G}_n(h)$ is close to $\mathbf{I}$, it can be expressed in terms of an element of Lie algebra $\Theta_n(h) \in \mathfrak{gl}$

$$\mathbf{G}_n(h) = \exp(h\, \boldsymbol{\Theta}_n(h)), \tag{111}$$

being a discrete solution of equation

$$\partial \boldsymbol{\Theta}_t = \mathbf{d}\text{exp}^{-1}_{\boldsymbol{\Theta}_t}(\boldsymbol{\Delta}_t)(\boldsymbol{\Upsilon}_t), \qquad t \geq 0, \qquad \boldsymbol{\Theta}_0 = \mathbf{0} \tag{112}$$

obtained in Sect. 4, which transforms the solution curve $\boldsymbol{\Upsilon}_t$ on $Sym^+\,(GL)$ locally around any point $\boldsymbol{\Upsilon}_0$ to a corresponding curve $\boldsymbol{\Theta}_t$ on the auxiliary vector space $sym\,(\mathfrak{gl})$ through $\mathbf{0}$. Since Lie algebras are linear vector spaces, we can then employ classical numerical methods, such as the Runge–Kutta methods. Moreover, working in Lie algebras instead of Lie groups enables to employ some very technical tools from Lie-algebra theory, such as the Lie bracket operation (24) that can also be used to great effect in reducing the cost of numerical approximations.

As a result, to solve (104), we set $\boldsymbol{\Upsilon}_0 \equiv \boldsymbol{\Upsilon}_n$ – the last approximation to $\boldsymbol{\Upsilon}(t_n)$, which will correspond to $\boldsymbol{\Theta}_0 = \mathbf{0}$ in each time step. Then the Runge–Kutta procedure with time step h can be applied to solve the local initial value problem (112) in terms of the weighted average $\boldsymbol{\Theta}$ (114b), and by transforming the result back via (114a) to the primary homogeneous space we eventually obtain an update $\boldsymbol{\Upsilon}_{n+1}$ – the approximation to $\boldsymbol{\Upsilon}(t_{n+1})$. This procedure is called the Runge–Kutta–Munthe-Kaas (RKMK) method [21].

Using the Butcher tableau, one step of the RKMK method consists in solving following system of equations for δ_l

$$\left.\begin{aligned} \boldsymbol{\Delta}_k &= \boldsymbol{\Delta}(t_n + c_k h, \Phi(\boldsymbol{\Upsilon}_n)(\exp(h\boldsymbol{\Theta}_k))) \\ \text{where } \boldsymbol{\Theta}_k &= \textstyle\sum_{l=0}^{\nu} a_{k,l}\delta_l \\ \delta_l &= \mathbf{d}\exp_{h\boldsymbol{\Theta}_l}^{-1}(\boldsymbol{\Delta}_l) \end{aligned}\right\} \quad k = 1, \ldots, \nu \tag{113}$$

then followed by the update

$$\boldsymbol{\Upsilon}_{n+1} = \Phi(\boldsymbol{\Upsilon}_n)(\exp(h\boldsymbol{\Theta})), \qquad \text{where} \quad \boldsymbol{\Theta} = \sum_{l=0}^{\nu} b_l\delta_l \,. \tag{114}$$

Here, the transformation Φ indexed by $\boldsymbol{\Upsilon}_n$ stands for the respective actions (108)-(110).

Further, smoothness of the transformation mapping $\exp$ ensures the correct order of the numerical approximation on the homogeneous manifold Sym^+, resp. on the group GL. To complete the algorithms, we need to provide practical means for computing $\mathbf{d}\exp_{\boldsymbol{\Theta}}^{-1}(\mathbf{D})$. For Eqs. (48) or (50), we may use expansion (82), truncated consistently to the order of the underlying Runge – Kutta scheme. The resulting RKMK scheme will obtain the same order as the underlying Runge – Kutta scheme while staying on the correct manifold all the time [17, 21]. For Eq. (49), the expression (92) provides a direct, finite-term algorithm, similar to the expression for the inverse of the trivialised tangent $\mathbf{d}\mathrm{cay}^{-1}$ on the Lie algebra $\mathfrak{o}$ of the orthogonal group O, when using Cayley coordinates $\mathrm{cay}: \mathfrak{o} \to O$ to an analysis of rotating rigid bodies [8, 17, 23].

6. Magnus and Fer Expansion Method

The Magnus and Fer expansion methods [16, 17, 18] are another methods that provide solution to the time-dependent *linear* equation evolving on the Lie group GL

$$\partial \mathbf{F}_t = \mathbf{L}_t \mathbf{F}_t = \rho_*(\mathbf{L}_t)(\mathbf{F}_t) \qquad \mathbf{F}|_{t=0} = \mathbf{F}_0 \qquad \mathbf{L} : \mathbb{R}^+ \to \mathfrak{gl}\,, \tag{115}$$

in terms of exponentials that include combinations of the coefficient $\mathbf{L}_t \subset \mathfrak{gl}$. Since the Lie algebra $\mathfrak{gl}$ is non-commutative, the solution

$$\mathbf{F}_t \neq \exp\left(\int_0^t \mathbf{L}_\xi \, d\xi\right)\mathbf{F}_0. \tag{116}$$

However, if we make a correction by adding a Lie algebra term to the argument in the exponential

$$\mathbf{F}_t = \exp\left(\int_0^t (\mathbf{L}_\xi + \mathbf{H}_\xi)\, d\xi\right)\mathbf{F}_0, \tag{117}$$

we can find the solution by Magnus expansion in terms of an infinite series of $\mathbf{H}_\xi$, or if this correction is realized by attaching a Lie group factor $\mathbf{G}_t$ behind the exponential

$$\mathbf{F}_t = \exp\left(\int_0^t \mathbf{L}_\xi\, d\xi\right)\mathbf{G}_t\mathbf{F}_0\,, \tag{118}$$

the Fer expansion gives the solution in terms of an iterative prescript for $\mathbf{G}_t$.

A wide range of applications of both Magnus and Fer expansion methods is rather broadly covered in [3]. Possibilities to solve also *non-linear* equations with these methods are developed in [6].

6.1. Magnus Expansion Method

Denoting $\boldsymbol{\Theta}_t = \mathbf{L}_t + \mathbf{H}_t$, then due to (80)-(82)

$$\partial\boldsymbol{\Theta}_t = \mathbf{dexp}^{-1}_{\boldsymbol{\Theta}_t}(\mathbf{L}_t) \equiv \sum_{j=0}^{\infty}\frac{B_j}{j\,!}\,[\boldsymbol{\Theta}_t, [\boldsymbol{\Theta}_t, [\ldots, [\boldsymbol{\Theta}_t, \mathbf{L}_t]\ldots]]] \tag{119}$$

is highly non-linear equation for $\boldsymbol{\Theta}_t$ on $\mathfrak{gl}$ with the initial condition $\boldsymbol{\Theta}_0 = \mathbf{0}$.

The main idea behind the Magnus expansion is to represent the solution as an infinite series

$$\boldsymbol{\Theta}_t = \sum_{n=1}^{\infty}\boldsymbol{\Theta}_n(t) \tag{120}$$

in a form that involves the coefficient $\mathbf{L}_t$ in repeated commutation and integration. Thus the truncated Magnus series also belongs to the Lie algebra, and so approximate solution still share important qualitative properties with the exact solution. The first three terms are

$$\boldsymbol{\Theta}_1(t) = \int_0^t \mathbf{L}(\xi)\, d\xi \tag{121}$$

$$\boldsymbol{\Theta}_2(t) = \frac{1}{2}\int_0^t d\xi_1 \int_0^{\xi_1} d\xi_2\, [\mathbf{L}(\xi_1), \mathbf{L}(\xi_2)] \tag{122}$$

$$\boldsymbol{\Theta}_3(t) = \frac{1}{6}\int_0^t d\xi_1 \int_0^{\xi_1} d\xi_2 \int_0^{\xi_2} d\xi_3\, ([\mathbf{L}(\xi_1), [\mathbf{L}(\xi_2), \mathbf{L}(\xi_3)]] + + [\mathbf{L}(\xi_3), [\mathbf{L}(\xi_2), \mathbf{L}(\xi_1)]]) \tag{123}$$

and, since the explicit expressions for higher-order terms in the Magnus series becomes difficult task, this three-terms approximation has been usually used.

The intrinsic and growing complexity of the Magnus expansion, however, allows for different schemes of construction of higher order terms, one of which is the recurrent scheme for $n \geq 2$

$$\boldsymbol{\Theta}_1(t) = \int_0^t \mathbf{L}(\xi)\, d\xi \tag{124}$$

$$\boldsymbol{\Theta}_n(t) = \sum_{j=1}^{n-1} \frac{B_j}{j\,!} \sum_{\substack{i_1 + \cdots + i_k = n-1 \\ i_1 \geq 1, \cdots, i_k \geq 1}} \int_0^t [\boldsymbol{\Theta}_{i_1}(\xi), [\cdots [\boldsymbol{\Theta}_{i_k}(\xi), \mathbf{L}(\xi)]] \cdots]\, d\xi. \tag{125}$$

The Magnus expansion coverges for values of t such that

$$\int_0^t \|\mathbf{L}(\xi)\|\, d\xi < \pi \tag{126}$$

in terms of the operator norm.

6.2. Fer Expansion Method

Rewriting the relation (118) as

$$\mathbf{F}(t) = \exp\left(\mathbf{G}_1(t)\right) \mathbf{F}_1(t) \tag{127}$$

with

$$\mathbf{G}_1(t) = \int_0^t \mathbf{L}(\xi)\, d\xi\,, \tag{128}$$

and substituting it into equation (119), then after some arrangements we arrive at

$$\partial \mathbf{F}_1(t) = \mathbf{L}_1(t)\,\mathbf{F}_1(t) \qquad \mathbf{F}_1(0) = \mathbf{Y}_0\,, \tag{129}$$

where

$$\mathbf{L}_1(t) = \exp(-\mathbf{G}_1(t))\,\mathbf{L}(t)\exp(\mathbf{G}_1(t)) - \\ - \int_0^t d\xi \exp(-\xi\mathbf{G}_1(\xi))\,\mathbf{L}(\xi)\exp(\xi\mathbf{G}_1(\xi))\,. \tag{130}$$

The above procedure can be repeated, and after n steps we arrive the following recursive scheme, known as the Fer expansion

$$\begin{aligned} \mathbf{F}(t) &= \exp(\mathbf{G}_1(t))\,\mathbf{F}_1(t) \\ \mathbf{F}_1(t) &= \exp(\mathbf{G}_2(t))\,\mathbf{F}_2(t) \\ &\vdots \\ \mathbf{F}_{n-1}(t) &= \exp(\mathbf{G}_n(t))\,\mathbf{F}_n(t)\,, \\ &\vdots \end{aligned}$$

where each individual term $\mathbf{F}_n(t)$ is a solution to the equation

$$\partial \mathbf{F}_n(t) = \mathbf{L}_n(t)\,\mathbf{F}_n(t) \qquad \mathbf{F}_n(0) = \mathbf{Y}_0 \qquad n = 1, 2, \cdots \tag{131}$$

with $\mathbf{L}_n(t)$ given by

$$\begin{aligned} \mathbf{L}_{n+1}(t) &= \exp(-\mathbf{G}_{n+1}(t))\,\mathbf{L}_n(t)\exp(\mathbf{G}_{n+1}(t)) - \\ &\qquad - \int_0^t d\xi \exp(-\xi\mathbf{G}_{n+1}(\xi))\,\mathbf{L}_n(\xi)\exp(\xi\mathbf{G}_{n+1}(\xi)) \\ &= \sum_{j=1}^{\infty} (-1)^j\,[\mathbf{G}_{n+1}(t), [\cdots[\mathbf{G}_{n+1}(t), \mathbf{L}_n(t)]]\cdots]\,. \end{aligned} \tag{132}$$

$$\mathbf{G}_{n+1}(t) = \int_0^t \mathbf{L}_n(\xi)d\xi \qquad \mathbf{L}_0(t) = \mathbf{L}(t) \qquad n = 0, 1, 2, \cdots \tag{133}$$

Operation of commutation in (133) is taken $j-$times. Truncation of the expansion after n steps yields an approximation to the exact solution $\mathbf{F}(t)$. The Fer expansion coverges for values of t such that

$$\int_0^t \|\mathbf{L}(\xi)\|\,d\xi < 0.8604065 \tag{134}$$

in terms of the operator norm.

Conclusion

We studied specific evolution equations of Lie type on the homogeneous space of all symmetric positive-definite matrices $Sym^+(3,\mathbb{R}) = GL(3,\mathbb{R})/O(3,\mathbb{R})$, and on the general linear group $GL(3,\mathbb{R})$, which originate in solid mechanics when describing finite deformations. The first case concerns the evolution of the right Cauchy – Green deformation tensor $\mathbf{C}_t$, the second one the evolution of the deformation gradient $\mathbf{F}_t$ on corresponding configuration spaces. Recognition of evolution equations of Lie type in solid mechanics is particularly possible due to its geometrical setting [11], which provides a novel approach to finite deformations, finally resulting in improving time-discrete integration schemes for incremental solutions to non-linear problems, cf. [13] versus [26, 27].

References

[1] Arvanitogeorgos, A.: *An Introduction to Lie Groups and the Geometry of Homogeneous Spaces. Student Mathematical Library* 22, AMS, Providence, Rhode Island, 2003.

[2] Bhatia, R.: *Positive Definite Matrices*. Princeton University Press, Princeton, 2007.

[3] Blanes, S., Casas, F., Oteo, J.A., Ros, J.: The Magnus expansion and some of its applications. *Phys. Rep.* 470(5-6), 151–238 (2009).

[4] Bridson, M.R., Haefliger, A.: *Metric Spaces of Non-Positive Curvature*. Springer, Berlin, 1999.

[5] Cariñena, J.F., de Lucas, J.: *Lie systems: theory, generalizations, and applications*. Dissertationes Mathematicae 479. Institute of Mathematics of the Polish Academy of Sciences, 2011.

[6] Casas, F., Iserles, A.: Explicit Magnus expansions for nonlinear equations. *J. Phys. A: Math. Gen.* 39(19), 5445–5461 (2006).

[7] Celledoni, E., Marthinsen, H., Owren, B.: An introduction to Lie group integrators – basics, new developments and applications. *J. Comput. Phys.* 257, 1040–1061 (2014).

[8] Celledoni, E., Owren, B.: Lie group methods for rigid body dynamics and time integration on manifolds. *Comput. Methods Appl. Mech. Eng.* 192(3-4), 421–438 (2003).

[9] Engø, K.: On the construction of geometric integrators in the RKMK class. *BIT* 40(1), 41–61 (2000).

[10] Faraut, J., Korányi J.: *Analysis on Symmetric Cones*. Oxford Science Publications, Oxford, 1994.

[11] Fiala, Z.: Geometrical setting of solid mechanics. *Ann. Phys.* 326(8), 1983–1997 (2011).

[12] Fiala, Z.: Evolution equation of Lie-type for finite deformations, time-discrete integration, and incremental methods. *Acta Mech.* 226(1), 17–35 (2015).

[13] Fiala, Z.: Geometry of finite deformations and time-incremental analysis. *Int. J. Nonlinear Mech.* 81, 230–244 (2016).

[14] Hairer, E., Nørsett, S.P., Wanner, G.: *Solving Ordinary Differential Equations I: Nonstiff Problems*. Springer, Berlin, 1993.

[15] Hairer, E., Lubich, Ch., Wanner, G.: *Geometric Numerical Integration. Structure-Preserving Algorithms for Ordinary Differential Equations*. Springer, Berlin, 2006.

[16] Iserles, A.: Expansions that grow on trees. *Not. AMS* 49(4), 430–440 (2002).

[17] Iserles, A., Munthe-Kaas, H.Z., Nørsett, S.P., Zanna, A.: Lie-group methods, *Acta Numer.* 9, 215–365 (2000).

[18] Iserles, A., Nørsett, S.P.: On the solution of linear differential equations in Lie groups. Phil. Trans. R. Soc. Lond. A 357, 983–1019 (1999).

[19] Isham, Ch.J.: *Modern Differential Geometry for Physicists*. World Scientific, Singapore, 2001.

[20] Jost, J.: *Riemannian Geometry and Geometric Analysis*. Springer, Berlin, 2002.

[21] Munthe-Kaas, H.Z.: High order Runge – Kutta methods on manifolds. *Appl. Numer. Math.* 29(1), 115–127 (1999).

[22] Munthe-Kaas, H.Z., Quispel, G.R.W., Zanna A.: Symmetric spaces and Lie triple systems in numerical analysis of differential equations. *BIT Numer. Math.* 54(1), 257–282 (2014).

[23] Mäkinen, J.: Critical study of Newmark-scheme on manifold of finite rotations. *Comput. Methods Appl. Mech. Eng.* 191(8-10), 817–828 (2001).

[24] Marsden, J.E., Hughes, T.J.R.: *Mathematical Foundations of Elasticity.* Dover, New York, 1993.

[25] Sattinger, D.H., Weaver, O.L.: *Lie Groups and Algebras with Applications to Physics, Geometry, and Mechanics.* Springer, Berlin, 1986.

[26] Simo, J.C.: *Numerical analysis and simulations of plasticity.* Handbook of numerical analysis VI, Ciarlet, P.G. & Lions, J.L. (eds.), 183–499. Elsevier, Amsterdam, 1998.

[27] Simo, J.C., Hughes, T.J.R.: *Computational Inelasticity.* Springer, Berlin, 1997.

[28] Trofimov, V.V.: Introduction to geometry of manifolds with symmetry. *Mathematics and Its Applications* 270, Springer, 1993.

In: Emerging Concepts in Evolution Equations ISBN: 978-1-53610-861-3
Editor: Carolyn Murphy, pp. 31-67

Chapter 2

NEW APPLICATIONS OF SYMMETRY ANALYSIS TO NONLINEAR EVOLUTION EQUATIONS

***Renat Zhdanov*[1,*] *and Qing Huang*[2]**
[1]BIO-key International, Eagan, MN, US
[2]Department of Mathematics, Northwest University, Xi'an, China

Abstract

In this chapter we present a review of our recent results on group analysis of nonlinear evolution equations in one spatial variable. The main emphasis is on the novel systematic approaches that expand the range of applications of the traditional Lie approach to evolution equations. We analyze two possible ways of expanding the applicability of Lie group methods: considering more general types of symmetries and applying the group approach to more general (fractional) evolution equations.

1. Introduction

Lie group analysis is universally recognized as a versatile and convenient tool for analyzing partial differential equations (PDEs). However, there is a necessary prerequisite for efficiently utilizing any group-theoretical method. Namely,

*E-mail address: Renat.Zhdanov@bio-key.com (Corresponding author).

the equation under study has to admit a nontrivial Lie group. This has led to numerous publications on the problem of group classification of nonlinear PDEs.

In the case when transformation involves dependent and independent variables only, it is called point transformation. For the more general case, when transformation includes first derivatives of the dependent variables, the term "contact transformation" has been adopted in the literature. Contact transformations play a significant role in analytical mechanics, where the canonical transformation is defined as the contact transformation which preserves the Poisson bracket.

Point group classification of the general second-order nonlinear evolution equations of the form

$$u_t = F(t, x, u, u_1, u_2), \tag{1}$$

where $u = u(t, x), u_t = \partial u/\partial t, u_i = \partial^i u/\partial x^i$ $(i \in \mathbb{N})$, and F is an arbitrary sufficiently smooth real-valued function with $F_{u_2} \neq 0$, has been extensively studied (see [59, 4, 60] and the references therein). In contrast, much less attention has been devoted to the contact symmetries of PDEs (1). Note that the class (1) includes a number of important fundamental equations of modern mathematical and theoretical physics, such as the heat, Fisher, Newell–Whitehead, and Burgers equations, to mention a few (see, e.g., [30, 39, 56]).

The notion of contact transformation within the context of differential equations (DEs) was first presented by Sophus Lie [27]. He obtained a number of classical results on contact symmetries of ordinary differential equations (ODEs). In addition, Lie described linear PDEs in two independent variables admitting contact symmetries [28]. Classification of contact symmetry groups of wave and related equations has been performed in [17]. Contact transformations of nonlinear hyperbolic type equations have been studied in [23, 36, 37, 38]. The paper [48] tackles the contact symmetry reduction of second-order PDEs. The inter-relationship between contact symmetries and conservation laws has been investigated in [24]. Recently, local conservation laws of second-order evolution equations have been classified up to contact equivalence [46], and the action of contact equivalence transformation on low-order conservation laws of evolution equations of an arbitrary order has been considered in [47]. The concept of the discrete transformation group has been modified to include contact transformations in [21].

Sokolov [51, 52] dealt with the evolutionary symmetries of the evolution equations

$$u_t = F(x, u, u_1, u_2, \cdots, u_n), \qquad n \geq 2, \tag{2}$$

and Magadeev [31] performed contact group classification of the equations of the form

$$u_t = F(t, x, u, u_1, u_2, \cdots, u_n), \qquad n \geq 2. \tag{3}$$

They obtained a number of nontrivial results on structure and dimension of contact symmetry algebras and described all possible realizations of symmetry algebras of the equations above by Lie vector fields over the field of complex numbers.

The method for computing algebras of contact symmetries developed by Sokolov relies heavily on the fact that an evolution equation by definition involves only the first-order derivative in the temporal variable, t. As a result, the transformation law for the temporal variable involves t only. In fact, the variable t enters contact symmetry almost like a parameter. This ensures that contact symmetries of an evolution PDE are closely related to contact symmetries of ordinary differential equations (ODEs) obtained by putting $u_t = 0$. Therefore the efficiency of Sokolov's approach relies on the well-known classification of contact symmetries of ODEs. One can look for finite-dimensional contact symmetries of (3) by considering linear combinations of the generators of contact symmetries of the corresponding ODEs and of the operator ∂_t with coefficients depending on t. In addition, the evolution equations that admit infinite contact symmetry are known to be linearizable. Evidently, such an approach has limited applicability beyond the class of parabolic types of PDEs.

In Sec. 2, we develop an alternative approach to classifying contact symmetries of PDEs. We apply this approach in order to perform group classification of second-order evolution equations (1) admitting contact symmetries. We solve completely the problem of constructing all inequivalent realizations of contact transformation groups containing semi-simple subgroups. Utilizing these results, we construct all Eqs. (1) that are invariant with respect to semi-simple contact transformation groups or groups containing semi-simple subgroup. Action of the group of contact transformations imposes stronger equivalence relation than that of the group of point transformations. By this very reason, we get fewer inequivalent realizations than we did for the case of point transformation groups in [4, 60]. Any symmetry algebra containing a semi-simple subalgebra obtained in [4, 60] is equivalent to one of the realizations derived in this paper with respect to a suitable contact transformation.

If one adopts the definition of the potential symmetry of an evolution PDE suggested by Bluman [6, 7], then the following assertion holds: any potential symmetry of this type admitted by an evolution equation in one spatial dimension can be mapped into a contact symmetry of a related evolution equation of the same order [19, 46, 47, 58]. Consequently, group classification of the class (1) admitting contact symmetries yields, as a by-product, an exhaustive description of PDEs (1) possessing potential symmetries. In other words, it provides nontrivial insight into the largely unexplored world of non-local symmetries of PDEs, since a potential symmetry is a special case of nonlocal symmetry.

Another recent trend in group analysis of differential equations is the modification of the classical Lie approach to make it applicable to fractional differential equations (FDEs).

Fractional calculus is based on the generalization of the standard differentiation and integration for an arbitrary non-integer order. It is as old as the classical calculus and goes back to the time when Leibnitz and Newton invented differential calculus. FDEs are generalizations of the differential equations (DEs) of integer order to non-integer one through the application of fractional calculus. It has been shown that new fractional-order models provide a more accurate description of nonlinear phenomena than their integer-order counterparts.

Analytical solutions are extremely helpful in providing deeper insight into the behavior of the system under consideration. Since FDEs appear more and more frequently in various research and engineering applications, such as rheology, visco-elasticity, biology, physics, and electrochemistry (see [2, 10, 25, 33, 41, 45, 57] and references therein), they attract considerable interest and have resulted in significant theoretical development (for example, see [25, 33, 45]). It is not surprising that there are no efficient general methods for solving these types of nonlinear equations. Various methods have been used to solve these equations, (such as Laplace transform [2], Fourier transform [33], variational iteration [55], finite difference [32], Adomian decomposition [34], homotopy perturbation [16, 35], generalized differential transform [40] and operational matrices [5]), all with different degrees of success.

Group analysis of PDEs is a well-established branch of the modern mathematical physics. The question is whether the Lie group theory can be efficiently used for analysis of FDEs just as it was for PDEs? The answer to this question is positive; however, the number of publication devoted to group analysis of FDEs and group properties of FDEs is rather low. That is why group properties of FDEs are much less understood, in striking contrast to PDEs.

Scaling transformation of the time fractional linear wave-diffusion equation and its group invariant solutions has been described in [8]. Using groups of scaling transformation, self-similar solutions to nonlinear fractional diffusion,

Burgers and KdV equations are obtained [12]. Lie symmetries of the fractional nonlinear anomalous diffusion equations are studied in [13, 14]. The problem of group analysis of fractional generalized Burgers and KdV equations is discussed in [49, 54]. Complete group classifications and symmetry reductions of the fractional fifth-order KdV types of equations are performed in [29]. Conservation laws and exact solutions of time-fractional Kompaneets equations are presented in [15]. The Lie group method to study time fractional evolution systems is developed in [20].

Time fractional PDEs are obtained from classical PDEs by replacing the time derivative with a fractional derivative.

In Sec. 3, we perform Lie group analysis, construct optimal systems and determine group invariant solutions of the time fractional Harry–Dym equation

$$\partial_t^\alpha u = u^3 u_{xxx}, \qquad (0 < \alpha < 1), \tag{4}$$

where $\partial_t^\alpha u$ is the Riemann–Liouville fractional derivative of order α with respect to the variable t.

2. Contact Group Classification of Nonlinear Evolution Equations: Semi-Simple Groups

To classify the family of PDEs (1) admitting contact symmetries, we modify the classification approach of [59] based on point symmetries in order to make it applicable beyond Lie point symmetries. Our group classification procedure is implemented in three major steps. First, we compute the most general group of contact symmetries that leaves the class (1) invariant. This yields the classifying PDEs for an infinitesimal symmetry generator and unknown function F. And what is more, one gets the explicit form of the equivalence group of contact transformations for (1). Second, using the well-known results on the classification of low-dimensional abstract Lie algebras (see, e.g., [3, 53]), we describe all inequivalent realizations of contact symmetry algebras by basis infinitesimal generators admitted by (1). Finally, we solve the classifying equations for the obtained realizations of Lie algebras, which yields the explicit forms of invariant equations. This approach enables exhaustive classification of second-order nonlinear evolution equations in one spatial dimension invariant under semi-simple groups of contact transformations.

If a finite-dimensional algebra L contains the radical N (the largest solvable ideal in L), then due to the Levi-Mal'cev theorem, there exists a semi-simple subalgebra S such that

$$L = S \ltimes N, \tag{5}$$

where S is the Levi factor. The relation (5) is called the Levi decomposition of L. Consequently, any Lie algebra falls into one of the following three categories: (i) semi-simple algebra, (ii) solvable algebra, and, (iii) semi-direct sum of solvable and semi-simple algebra.

Here we restrict our considerations to the cases when symmetry group is either semi-simple or has a nontrivial Levi factor.

2.1. Admissible Transformations of Evolution Equations

To keep the exposition self-contained, we summarize below the well known facts from the theory of contact symmetries of evolution equations. An interested reader can find further details in the monographs [1, 9, 22, 43].

It is a common knowledge that the most general contact transformation group admitted by (1) is generated by the infinitesimal operator

$$V_g = -g_{u_t}\partial_t - g_{u_1}\partial_x + (g - u_t g_{u_t} - u_1 g_{u_1})\partial_u + (g_x + u_1 g_u)\partial_{u_1} + (g_t + u_t g_u)\partial_{u_t}, \tag{6}$$

where the arbitrary real-valued smooth function $g = g(t, x, u, u_t, u_1)$ is called the generating function or the contact Hamiltonian for the vector field V_g. Since function g uniquely defines V_g, we usually use g as a short form of V_g throughout the section. V_g can also be represented in the equivalent form of Lie-Bäcklund vector field (LBVF)

$$g\partial_u + (D_x g)\partial_{u_1} + (D_t g)\partial_{u_t} + (D_x^2 g)\partial_{u_2} + \cdots, \tag{7}$$

where $D_x^{j+1} = D_x(D_x^j)$, and the symbols D_t and D_x stand for the total differentiation operators with respect to the variables t and x, respectively. Namely,

$$D_t = \partial_t + u_t\partial_u + u_{tt}\partial_{u_t} + u_{xt}\partial_{u_1} + \cdots,$$

$$D_x = \partial_x + u_1\partial_u + u_{tx}\partial_{u_t} + u_2\partial_{u_1} + \cdots.$$

Furthermore, provided the generating function g is linear in u_t and u_1

$$g = \eta(t, x, u) - \tau(t, x, u)u_t - \xi(t, x, u)u_1,$$

both the vector field (6) and LBVF (7) are equivalent to the vector field $\tau(t,x,u)\partial_t + \xi(t,x,u)\partial_x + \eta(t,x,u)\partial_u$, which generates Lie group of point transformations. If g does not explicitly depend on t and u_t, it is called evolutionary symmetry [52].

As the form of (7) is uniquely determined by the corresponding function g, it is convenient to use the shorthand notation $g\partial_u$ instead of its full version (7). The class (1) is invariant under the LBVF $g\partial_u$ or, equivalently, under the vector field (6) if and only if the condition

$$[gF_u + D_x gF_{u_1} + D_x^2 gF_{u_2} - D_t g]|_{u_t=F} = 0$$

holds.

Inserting $u_t = F$ and its differential consequences into the invariance criterion and splitting the obtained relation with respect to the functionally independent variables yields an over-determined system of linear PDEs for unknown functions g and F. Solving the system, we arrive at the following assertion.

Lemma 2.1. *The most general contact symmetry admitted by* (1) *is of the form*

$$g(t,x,u,u_t,u_1) = \alpha(t)u_t + G(t,x,u,u_1), \tag{8}$$

where α and G are real-valued functions satisfying the classifying equation

$$\begin{aligned} &- G_{u_1}F_u u_1 + F_{u_2}G_u u_2 + 2F_{u_2}G_{u_1x}u_2 + 2F_{u_2}G_{ux}u_1 + F_{u_1}G_u u_1 \\ &+ F_{u_2}G_{uu}u_1^2 + F_{u_2}G_{u_1u_1}u_2^2 + 2F_{u_2}u_1G_{u_1u}u_2 + F_uG + F_{u_2}G_{xx} \\ &- G_{u_1}F_x + F_{u_1}G_x - \alpha F_t - FG_u - \alpha_t F - G_t = 0. \end{aligned} \tag{9}$$

Hence the problem of classifying the contact symmetries of the class (1) reduces to constructing all possible solutions of (9). The difficulty, however, is that (9) is an under-determined system of one PDE for three unknown functions α, G and F. To proceed with the classification, we need additional information on the structure of α, G and F. This information is provided by the classical results on the structure of abstract low dimensional Lie algebras.

Following the approach of [59], we have succeeded in obtaining constraints on the admissible forms of the functions α and G. This enabled constructing all inequivalent contact symmetries of the form (8). Inserting the obtained functions α and G into classifying equations and solving them, we derive the explicit forms of F.

We now look for contact transformations which map the class (1) onto itself and form the contact equivalence group of (1). The direct computation shows that the contact equivalence group admitted by the class (1) reads as

$$\tilde{t} = T(t), \quad \tilde{x} = X(t, x, u, u_1), \quad \tilde{u} = U(t, x, u, u_1), \tag{10}$$

where the functions T, X and U satisfy the regularity

$$T_t \neq 0, \qquad \text{rank} \begin{pmatrix} X_x & X_u & X_{u_x} \\ U_x & U_u & U_{u_x} \end{pmatrix} = 2$$

and the tangency conditions

$$X_{u_1}(u_1 U_u + U_x) = U_{u_1}(u_1 X_u + X_x). \tag{11}$$

Consider now the action of a contact transformation on the generating function and the corresponding equation. One can verify by straightforward computation that a contact transformation (10) maps the generating function

$$\tilde{g}(\tilde{t}, \tilde{x}, \tilde{u}, \tilde{u}_{\tilde{t}}, \tilde{u}_{\tilde{x}}) = \tilde{\alpha}(\tilde{t})\tilde{u}_{\tilde{t}} + \tilde{G}(\tilde{t}, \tilde{x}, \tilde{u}, \tilde{u}_{\tilde{x}})$$

of contact symmetry of the equation

$$\tilde{u}_{\tilde{t}} = \tilde{F}(\tilde{t}, \tilde{x}, \tilde{u}, \tilde{u}_{\tilde{x}}, \tilde{u}_{\tilde{x}\tilde{x}})$$

into the generating function

$$\begin{aligned} g(t, x, u, u_t, u_1) \quad = \quad & \frac{\tilde{\alpha}(T)}{\dot{T}} u_t + \frac{D(X)}{J} \left[\frac{D(X)U_t - D(U)X_t}{D(X)\dot{T}} \tilde{\alpha}(T) \right. \\ & \left. + \tilde{G}(T, X, U, \tilde{u}_{\tilde{x}}) \right], \end{aligned} \tag{12}$$

and the equation reduces to

$$u_t = \frac{D(X)}{J} \left[\dot{T}\tilde{F}(T, X, U, \tilde{u}_{\tilde{x}}, \tilde{u}_{\tilde{x}\tilde{x}}) + X_t \tilde{u}_{\tilde{x}} - U_t \right].$$

Here $D = \partial_x + \sum\limits_{i=0}^{\infty} u_{i+1}\partial_{u_i}$, $D^2 = D \circ D$, $J = D(X)U_u - D(U)X_u$ and

$$\tilde{u}_{\tilde{x}} = \frac{D(U)}{D(X)}, \quad \tilde{u}_{\tilde{x}\tilde{x}} = \frac{D(X)D^2(U) - D(U)D^2(X)}{D^3(X)}.$$

The formulas above reflect the simple fact that there exist infinitely many equivalent realizations of a contact symmetry. Choosing appropriately the transformation functions T, X and U in (10), one can reduce g to the canonical form. If a contact symmetry can be mapped into a point symmetry by a suitable contact transformation, it is called reducible symmetry. Reducibility is an important concept in the theory of contact transformation groups and Lie algebras of contact vector fields [9]. Given a choice, inequivalent representatives of reducible contact realizations should be chosen among point realizations of the corresponding group or algebra. So, whenever possible, the generating function g should be transformed to become linear in u_t and u_1.

Lemma 2.2. ([51, 52, 31]) Given an arbitrary generating function (8), there exists a contact transformation (10) reducing it to either 1 or u_t.

Solving the corresponding classifying equations yields the forms of all inequivalent equations belonging to (1) admitting one-dimensional Lie algebras of contact symmetries.

Theorem 2.1. *There exist only two inequivalent equations of the form* (1)*, which are invariant under one-parameter contact symmetry groups*

$$A_1^1 = \langle u_t \rangle : \quad u_t = F(x, u, u_1, u_2),$$

$$A_1^2 = \langle 1 \rangle : \quad u_t = F(t, x, u_1, u_2).$$

Next, we give the definition of the Lie bracket for generating functions f and g, which correspond to the commutation relation of the respective LBVFs. Denoting

$$f_*(g) = \partial_{u_t} f D_t g + \partial_u f g + \partial_{u_1} f D_x g,$$

we define the Lie bracket as follows

$$[f, g] = g_*(f) - f_*(g).$$

Computing the right-hand side of the above formula yields

$$\begin{aligned}[f, g] = {} & (\partial_{u_t} g \partial_u f - \partial_{u_t} f \partial_u g) u_t + (\partial_{u_1} g \partial_u f - \partial_{u_1} f \partial_u g) u_1 \\ & + \partial_{u_t} g \partial_t f - \partial_{u_t} f \partial_t g + \partial_{u_1} g \partial_x f - \partial_{u_1} f \partial_x g + f \partial_u g - g \partial_u f.\end{aligned} \tag{13}$$

Hence we find that $[f, g]$ does not depend on u_2, u_{tx} or u_{tt}. Consequently, contact symmetries of the class (1) form a Lie algebra with respect to the so defined Lie bracket. It is straightforward to verify that $[V_f, V_g] = V_{[f,g]}$, where $[V_f, V_g]$ is the standard Lie bracket and $[f, g]$ is given by (13).

In what follows, we will concentrate on the class of PDEs (1) invariant under semi-simple algebras and algebras having nontrivial Levi factors.

2.2. Equations Invariant under Semi-Simple Lie Algebras

Now we analyze equations from the class (1) whose contact symmetry algebras are semi-simple.

The dimension of a semi-simple Lie algebra is greater or equal to three. There are two non-isomorphic three-dimensional semi-simple Lie algebras, namely,

$$\mathfrak{sl}(2,\mathbb{R}): \quad [g_1, g_2] = g_1,\ [g_1, g_3] = 2g_2,\ [g_2, g_3] = g_3;$$
$$\mathfrak{so}(3): \quad [g_1, g_2] = g_3,\ [g_1, g_3] = -g_2,\ [g_2, g_3] = g_1.$$

The following assertion holds.

Theorem 2.2. *There are at most three inequivalent contact realizations of the Lie algebra $\mathfrak{sl}(2,\mathbb{R})$. The realizations and the corresponding invariant equations are as follows*

$$\mathfrak{sl}^1(2,\mathbb{R}) = \langle 1, u + xu_1, u^2 + 2xuu_1\rangle : \ u_t = xu_1 F\left(t, \frac{3u_1 + 2xu_2}{x^4 u_1^3}\right),$$

$$\mathfrak{sl}^2(2,\mathbb{R}) = \langle 1, u + xu_1, u^2 + 2xuu_1 + u_1\rangle :$$
$$u_t = (u_1 - x^2u_1^2)^{\frac{1}{2}} F\left(t, \frac{u_2 + 6xu_1^2 - 4x^3u_1^3}{(u_1 - x^2u_1^2)^{\frac{3}{2}}}\right),$$

$$\mathfrak{sl}^3(2,\mathbb{R}) = \langle u_t, -tu_t + xu_1, t^2u_t - 2txu_1 - u_1\rangle : \ u_t = -x^2u_1 + \frac{1}{u_1}F\left(u, \frac{u_2}{u_1^2}\right).$$

Moreover, any contact realization of the algebra $\mathfrak{so}(3)$ is equivalent to

$$\mathfrak{so}^1(3) = \langle 1, \tan x \sin u - u_1 \cos u, \tan x \cos u + u_1 \sin u\rangle.$$

The most general $\mathfrak{so}^1(3)$ invariant equation in the class (1) has the form

$$u_t = (\sec^2 x + u_1^2)^{\frac{1}{2}} F\left(t, \frac{u_2 \cos x - (2 + u_1^2\cos^2 x)u_1 \sin x}{(1 + u_1^2\cos^2 x)^{\frac{3}{2}}}\right).$$

Proof. We begin by analyzing inequivalent realizations of the algebra $\mathfrak{sl}(2, \mathbb{R})$. Choosing the basis operators g_i $(i = 1, 2, 3)$ in the general form (8), inserting them into the commutation relations of $\mathfrak{sl}(2, \mathbb{R})$, and solving the equations obtained yields all possible realizations of the algebra under study.

In view of Lemma 2.2, we can assume, without any loss of generality, that one of the basis elements, say g_1, can be reduced to 1 or u_t. We consider the case $g_1 = 1$ in full detail.

Let $g_1 = 1$ and g_2, g_3 be of the form (8). Inserting g_2, g_3 into the first two commutation relations from $\mathfrak{sl}(2, \mathbb{R})$ gives

$$g_2 = u + \phi(t, x, u_1).$$

Hereafter ϕ is an arbitrary real-valued smooth function of indicated variables.

To get all possible forms of the function g_2, we need to use contact transformations (10) preserving the basis element $g_1 = 1$, namely,

$$\tilde{g_1} = 1 \rightarrow g_1 = \frac{D(X)}{J} = 1.$$

Hence, we conclude that

$$\frac{D(X)}{D(X)U_u - D(U)X_u} = 1 \tag{14}$$

and

$$X_{u_1}D(U) = U_{u_1}D(X). \tag{15}$$

Eq. (14) can be rewritten as follows

$$1 = U_u - \frac{D(U)}{D(X)}X_u. \tag{16}$$

Consider the cases $X_{u_1} \neq 0$ and $X_{u_1} = 0$ separately.

Case 1. If $X_{u_1} \neq 0$, then it follows from (15) that $D(U)/D(X) = U_{u_1}/X_{u_1}$. Inserting this expression into (16) gives the first order linear PDE

$$X_{u_1} = X_{u_1}U_u - X_uU_{u_1}$$

for U. Rewriting it in the form

$$\frac{\mathrm{d}t}{0} = \frac{\mathrm{d}x}{0} = \frac{\mathrm{d}u}{X_{u_1}} = \frac{\mathrm{d}u_1}{-X_u} = \frac{\mathrm{d}U}{X_{u_1}}$$

and solving the obtained equation, we get the functionally-independent first integrals t, x, X and $U - u$.

Consequently, we can choose the contact transformation in the form

$$\tilde{x} = X(t, x, u, u_1), \quad \tilde{u} = u + Y(t, x, X). \tag{17}$$

In view of (17), tangency condition (15) leads to

$$X_{u_1}(u_1 + Y_x) = 0,$$

whence

$$Y_x = -u_1. \tag{18}$$

Thus the most general contact transformation leaving $g_1 = 1$ invariant is given by (17), where Y satisfies the condition (18).

Case 2. If $X_{u_1} = 0$, we have $D(X) \neq 0$ (otherwise $X = X(t)$ and the non-degeneracy assumption does not hold). Consequently, (15) implies that $U_{u_1} = 0$ and contact transformation (10) turns out to be a point transformation.

It follows from (16) that the relation

$$X_x + u_1 X_u = X_x U_u - X_u U_x$$

holds. As functions X and U are independent of u_1, we have $X_u = 0$. Hence $X = X(t, x)$ and $U_u = 1$. Then the most general point transformation that preserves $g_1 = 1$ is of the form

$$\tilde{t} = T(t), \quad \tilde{x} = X(t, x), \quad \tilde{u} = u + Y(t, x), \tag{19}$$

where $X_x \neq 0$.

Consider now the action of contact transformation (17) on g_2,

$$\tilde{g}_2 = \tilde{u} + \phi(\tilde{t}, \tilde{x}, \tilde{u}_{\tilde{x}}) \to g_2 = u + Y(t, x, X) + \phi(T, X, Y_X).$$

By choosing Y which satisfy the compatible system of PDEs

$$Y(t, x, X) + \phi(T, X, Y_X) = 0$$

and (18) we arrive at $g_2 = u$. The commutation relations imply that $g_3 = u^2 + \psi(t, x)u_1^2$, where ψ is an arbitrary function. Applying the transformation

$$\tilde{t} = T(t), \quad \tilde{x} = X(t, x), \quad \tilde{u} = u, \quad X_x \neq 0,$$

that preserves 1 and u, we arrive at

$$\tilde{g}_3 = \tilde{u}^2 + \psi(\tilde{t}, \tilde{x})\tilde{u}_{\tilde{x}}^2 \to g_3 = u^2 + \frac{\psi(T, X)}{X_x^2} u_1^2.$$

With properly chosen X and U, the function ψ/X_x^2 reduces to 0, 1 or -1. As a result, we obtain the following three inequivalent contact realizations of $\mathfrak{sl}(2, \mathbb{R})$:

$$\langle 1, u, u^2 \rangle, \quad \langle 1, u, u^2 - u_1^2 \rangle, \quad \langle 1, u, u^2 + u_1^2 \rangle.$$

In view of the reducibility requirement, the inequivalent representatives of reducible contact realizations should be chosen among point realizations, whenever possible. Let us analyze reducibility of the last two realizations. Applying the contact transformation

$$\tilde{t} = t, \quad \tilde{x} = \ln |x^2 u_1|, \quad \tilde{u} = u + x u_1,$$

to the realization $\langle 1, u, u^2 - u_1^2 \rangle$, we get

$$\langle 1, u + x u_1, u^2 + 2xuu_1 \rangle.$$

Analogously, the realization $\langle 1, u, u^2 + u_1^2 \rangle$ can be reduced to $\langle 1, u + xu_1, u^2 + 2xuu_1 + u_1 \rangle$ by the contact transformation

$$\tilde{t} = t, \quad \tilde{x} = \arcsin(2x^2 u_1 - 1), \quad \tilde{u} = u + x u_1.$$

Turn now to the case $g_1 = u_t$. Analysis of this case yields two more inequivalent realizations of the algebra $\mathfrak{sl}(2, \mathbb{R})$, namely, $\langle u_t, -tu_t + xu_1, t^2 u_t - 2txu_1 - u_1 \rangle$ and $\langle u_t, -tu_t, t^2 u_t \rangle$.

The algebra $\mathfrak{so}(3)$ is handled in a similar fashion. We get the realization $\mathfrak{so}^1(3)$ listed in the theorem.

To complete the proof, we need to construct the corresponding invariant equations. Inserting the realizations obtained above into the classifying equation (9), we verify that the algebras $\langle 1, u, u^2 \rangle$ and $\langle u_t, -tu_t, t^2 u_t \rangle$ cannot be symmetry algebras of (1). The remaining realizations give rise to the invariant equations given above. □

Remark 1. The realizations $\mathfrak{sl}^1(2, \mathbb{R})$ and $\mathfrak{sl}^2(2, \mathbb{R})$ are equivalent when considered over the field of complex numbers. Replacing the underlying field of real numbers by complex numbers thus leads to stronger equivalence relation and fewer inequivalent cases. Consequently, there are two inequivalent realizations of $\mathfrak{sl}(2,\mathbb{C})$ evolution symmetries

$$\langle 1, u, u^2\rangle, \qquad \langle 1, u, u^2 - u_1^2\rangle,$$

and another realizations of $\mathfrak{sl}(2,\mathbb{C})$ algebra of type III [31]. These results coincide with the realizations obtained in [31, 51, 52].

Remark 2. The algebra $\mathfrak{so}(3)$ over the complex field is isomorphic to $\mathfrak{sl}(2,\mathbb{C})$. So that the realization $\mathfrak{so}^1(3)$ is equivalent to the realization $\mathfrak{sl}^1(2,\mathbb{R})$.

Theorem 2.3. *Any equation of the form* (1) *invariant under a semi-simple Lie algebra of contact symmetries is equivalent to one of the PDEs listed in Theorem 2.2.*

Proof is carried out in the same way as it has been done in our paper [60], where we studied the less general case of Lie point transformations.

2.3. Equations Invariant under the Algebras Having Nontrivial Levi Factor

In what follows, we classify Eqs. (1) admitting the contact symmetry algebras having nontrivial Levi decomposition. Namely, we consider contact symmetry algebras that are represented as either direct or truly semi-direct sums of semi-simple and solvable Lie algebras. These two cases are to be considered separately.

2.3.1. Direct Sums of Semi-Simple and Solvable Lie Algebras

It suffices to consider only the direct sums of one of the four realizations of semi-simple algebras given in Theorem 2.2 and solvable algebras. Take $\mathfrak{sl}^1(2,\mathbb{R})$ as an example. We look for all possible extensions of the realization $\mathfrak{sl}^1(2,\mathbb{R})$ by functions (8) which commute with its basis functions. Analysis of the commutation conditions gives the most general form of the generating function

$$g = \alpha(t)u_t + \phi(t)xu_1. \tag{20}$$

Now we need to construct all possible solvable Lie algebras having the generating function (20). Skipping the intermediate computations, we formulate the final result: any solvable Lie algebra with generators (20) is isomorphic to one-dimensional algebras $\langle u_t\rangle$, $\langle \phi(t)xu_1\rangle$ with $\dot{\phi} \neq 0$ or the two-dimensional non-Abelian algebra $\langle u_t, -tu_t + \lambda x u_1\rangle$. Consequently, the list of all possible extensions of $\mathfrak{sl}^1(2,\mathbb{R})$ consists of three inequivalent algebras

$\mathfrak{sl}^1(2,\mathbb{R}) \oplus \langle u_t\rangle$:

$$u_t = xu_1 F\left(\frac{3u_1 + 2xu_2}{x^4 u_1^3}\right),$$

$\mathfrak{sl}^1(2,\mathbb{R}) \oplus \langle \phi(t) x u_1\rangle$, $\dot{\phi} \neq 0$:

$$u_t = \frac{x\dot{\phi}u_1}{2\phi}\ln\left|\frac{3u_1 + 2xu_2}{x^4 u_1^3}\right| + xu_1 F(t),$$

$\mathfrak{sl}^1(2,\mathbb{R}) \oplus \langle u_t, -tu_t + \lambda x u_1\rangle$, $\lambda \neq 0$:

$$u_t = Cx^{\frac{2+\lambda}{\lambda}} u_1^{\frac{3+2\lambda}{2\lambda}} (3u_1 + 2xu_2)^{-\frac{1}{2\lambda}}.$$

The extensions of $\mathfrak{sl}^2(2,\mathbb{R})$ and $\mathfrak{sl}^3(2,\mathbb{R})$ are obtained in a similar way.

$\mathfrak{sl}^2(2,\mathbb{R}) \oplus \langle u_t\rangle$:

$$u_t = (u_1 - x^2u_1^2)^{\frac{1}{2}} F\left(\frac{u_2 + 6xu_1^2 - 4x^3u_1^3}{(u_1 - x^2u_1^2)^{\frac{3}{2}}}\right),$$

$\mathfrak{sl}^2(2,\mathbb{R}) \oplus \langle \frac{\phi(t)}{x}(x^2u_1 - x^4u_1^2)^{\frac{1}{2}}\rangle$, $\dot{\phi} \neq 0$:

$$u_t = \frac{u_1^{\frac{1}{2}}(x^2u_1 - 1)^{\frac{1}{2}}(-u_1^3 + 3u_1^4x^2 - 3u_1^5x^4 + u_1^6x^6)^{\frac{1}{2}}}{u_1^{\frac{3}{2}}(x^2u_1 - 1)^{\frac{3}{2}}} F(t) + \frac{\dot{\phi}u_1^2(x^2u_1 - 1)^2 \operatorname{arctanh}\frac{u_2 + 6u_1^2x - 4x^3u_1^3}{4u_1^{\frac{3}{2}}(x^2u_1 - 1)^{\frac{3}{2}}}}{\phi u_1^{\frac{3}{2}}(x^2u_1 - 1)^{\frac{3}{2}}},$$

$\mathfrak{sl}^2(2,\mathbb{R}) \oplus \langle u_t, -tu_t + \frac{\lambda}{x}(x^2u_1 - x^4u_1^2)^{\frac{1}{2}}\rangle$, $\lambda \neq 0$:

$$u_t = Cu_1^{\frac{1}{2}}(x^2u_1 - 1)^{\frac{1}{2}} \exp\left(\frac{(x^2u_1 - x^4u_1^2)^{\frac{3}{2}}}{\lambda x^3 u_1^{\frac{3}{2}}(x^2u_1 - 1)^{\frac{3}{2}}} \operatorname{arctanh}\frac{-u_2 - 6u_1^2x + 4x^3u_1^3}{4u_1^{\frac{3}{2}}(x^2u_1 - 1)^{\frac{3}{2}}}\right),$$

$\mathfrak{sl}^3(2,\mathbb{R}) \oplus \langle 1 \rangle$:

$$u_t = -x^2 u_1 + \frac{1}{u_1} F\left(\frac{u_2}{u_1^2}\right),$$

$\mathfrak{sl}^3(2,\mathbb{R}) \oplus \langle 1, u \rangle$:

$$u_t = -x^2 u_1 + C\frac{u_1^3}{u_2^2}.$$

Analogously, the extension $\mathfrak{so}^1(3) \oplus \langle u_t \rangle$ yields one more invariant equation

$$u_t = (\sec^2 x + u_1^2)^{\frac{1}{2}} F\left(\frac{u_2 \cos x - (2 + u_1^2 \cos^2 x) u_1 \sin x}{(1 + u_1^2 \cos^2 x)^{\frac{3}{2}}}\right).$$

Note that the above symmetry algebras are maximal invariance algebras, in Lie's sense, of the corresponding equations provided the function F and constant C are arbitrary.

When considering the direct sums of semi-simple and solvable Lie algebras over the complex field, we only need to take into account the extensions of the realizations $\mathfrak{sl}^1(2,\mathbb{R})$ and $\mathfrak{sl}^3(2,\mathbb{R})$, since $\mathfrak{sl}^1(2,\mathbb{R})$, $\mathfrak{sl}^2(2,\mathbb{R})$ and $\mathfrak{so}^1(3)$ are equivalent. In this way we obtain all of the above realizations except for those associated with $\mathfrak{sl}^2(2,\mathbb{R})$ and $\mathfrak{so}^1(3)$.

2.3.2. Semi-Direct Sums of Semi-Simple and Solvable Lie Algebras

To classify PDEs (1), whose invariance algebras are isomorphic to semi-direct sums of semi-simple and solvable Lie algebras, we apply the following two-step approach.

First, we utilize the results of classification of low dimensional Lie algebras, which are semi-direct sum of Levi factor and solvable radical and are obtained in [53]. Then we describe all inequivalent equations within the class (1) admitting those algebras. Without loss of generality, we can restrict our considerations to the Lie algebras having Levi decomposition $\mathfrak{sl}(2,\mathbb{R}) \ltimes A_{2.1}$ and $\mathfrak{so}(3) \ltimes A_{3.1}$, where $A_{2.1}$ and $A_{3.1}$ are two- and three-dimensional Abelian algebras respectively [53].

Next, applying the Ovsiannikov's approach [44], we complete the group classification of PDEs (1) containing arbitrary functions of one variable or arbitrary constants.

Consider first the algebra $\mathfrak{sl}(2,\mathbb{R}) \ltimes A_{2.1}$. Choosing $\mathfrak{sl}(2,\mathbb{R}) = \langle g_1, g_2, g_3\rangle$ and $A_{2.1} = \langle g_4, g_5\rangle$ yields the following nonzero commutation relations:

$$[g_1, g_2] = g_1, \quad [g_1, g_3] = 2g_2, \quad [g_2, g_3] = g_3, \quad [g_1, g_5] = g_4,$$
$$[g_2, g_4] = -\tfrac{1}{2}g_4, \quad [g_2, g_5] = \tfrac{1}{2}g_5, \quad [g_3, g_4] = -g_5.$$

Here we provide the full details of the analysis of the algebra $\mathfrak{sl}^1(2,\mathbb{R}) \ltimes A_{2.1}$. Other cases are treated in a similar way, and we only present the final results. Let $g_1 = 1,\ g_2 = u + xu_1,\ g_3 = u^2 + 2xuu_1$, and $g_4,\ g_5$ be of the general form (8). Inserting these expressions into the commutation relations above and solving the obtained equations yields

$$g_4 = \phi(t)|x|^{-\frac{1}{2}}, \quad g_5 = \phi(t)|x|^{-\frac{1}{2}}(u + 2xu_1).$$

Performing the contact transformation

$$\tilde{t} = T(t), \quad \tilde{x} = xY(t), \quad \tilde{u} = u,$$

which preserves $\mathfrak{sl}^1(2,\mathbb{R})$, we have

$$g_4 = \phi(T)|Y(t)|^{-\frac{1}{2}}|x|^{-\frac{1}{2}}, \quad g_5 = \phi(T)|Y(t)|^{-\frac{1}{2}}|x|^{-\frac{1}{2}}(u + 2xu_1).$$

Choosing $T = t$ and $Y = \phi^2(t)$, this $A_{2.1}$ algebra reduces to $\langle |x|^{-\frac{1}{2}}, |x|^{-\frac{1}{2}}(u + 2xu_1)\rangle$. Consequently, any extension of the algebra $\mathfrak{sl}^1(2,\mathbb{R})$ is equivalent to

$$\mathfrak{sl}^1(2,\mathbb{R}) \ltimes \langle |x|^{-\frac{1}{2}}, |x|^{-\frac{1}{2}}(u + 2xu_1)\rangle.$$

The corresponding invariant equation takes the form

$$u_t = \frac{(3u_1 + 2xu_2)^{\frac{1}{3}}}{x^{\frac{1}{3}}}F(t). \tag{21}$$

However, the five-dimensional Lie algebra is not the maximal one admitted by (21). In order to determine its maximal symmetry algebra, we simplify the equation under study. By making the change of variables

$$\tilde{t} = \int F(t)\mathrm{d}t, \quad \tilde{x} = x, \quad \tilde{u} = u,$$

Eq. (21) is transformed to

$$u_t = \frac{(3u_1 + 2xu_2)^{\frac{1}{3}}}{x^{\frac{1}{3}}}.$$

Applying the Lie infinitesimal algorithm we prove that the maximal invariance algebra of the transformed equation is the seven-dimensional Lie algebra

$$\mathfrak{sl}^1(2,\mathbb{R}) \ltimes \langle |x|^{-\frac{1}{2}}, |x|^{-\frac{1}{2}}(u + 2xu_1), u_t, -\frac{4}{3}tu_t - 2xu_1\rangle,$$

which is isomorphic to $\mathfrak{sl}(2,\mathbb{R}) \ltimes A_{4.5}$. Here $A_{4.5}$ is the four-dimensional decomposable Lie algebra with $q = 1$ and $p = 4/3$ given in [4].

Analogous analysis shows that $\mathfrak{sl}^2(2,\mathbb{R})$ cannot be extended up to an invariance algebra of (1) isomorphic to $\mathfrak{sl}(2,\mathbb{R}) \ltimes A_{2.1}$. The algebra $\mathfrak{sl}^3(2,\mathbb{R})$ admits the realization

$$\mathfrak{sl}^3(2,\mathbb{R}) \ltimes \langle |u_x|^{-\frac{1}{2}}(xu_1 + u), -|u_x|^{-\frac{1}{2}}((tx + 1)u_1 + tu)\rangle,$$

which is the invariance algebra of PDE

$$u_t = -2\lambda\frac{u_1}{u_2} + \frac{u^2 + \lambda u}{u_1} - x^2u_1, \quad \lambda \neq 0.$$

The maximal invariance algebra of the above equation is five-dimensional.

Turn now to the algebra $\mathfrak{so}(3) \ltimes A_{3.1}$. Let $\mathfrak{so}(3) = \langle g_1, g_2, g_3\rangle$ and $A_{3.1} = \langle g_4, g_5, g_6\rangle$, where g_i $(i = 1, 2, \cdots, 5)$, satisfy the nonzero commutation relations

$$[g_1, g_5] = g_6, \quad [g_1, g_6] = -g_5, \quad [g_2, g_4] = -g_6,$$
$$[g_2, g_6] = g_4, \quad [g_3, g_4] = g_5, \quad [g_3, g_5] = -g_4.$$

Letting $g_1 = 1$, $g_2 = \tan x \sin u - u_1 \cos u$, $g_3 = \tan x \cos u + u_1 \sin u$, g_4, g_5, g_6 be of the general form (8), and inserting the expressions for g_i into the commutation relations above, leads to the trivial generators $g_4 = g_5 = g_6 = 0$. By this reason PDEs (1) cannot admit a symmetry algebra that has Levi decomposition $\mathfrak{so}(3) \ltimes A_{3.1}$.

Note that the analysis of realizations of the semi-direct sums of semi-simple and solvable Lie algebras over the complex field yields the same realizations as those obtained above.

3. Symmetries and Exact Solutions of the Time Fractional Harry-Dym Equation with the Riemann–Liouville Derivative

In this section, we perform group analysis of the time fractional Harry–Dym equation with the Riemann–Liouville derivative by computing its maximal symmetry group and the corresponding optimal system of subgroups. Using these results we perform similarity reductions of the fractional Harry–Dym equation and construct its explicit group invariant solutions.

3.1. Preliminaries

There are several definitions of the fractional derivative operators which are, generally speaking, not equivalent to each other. We adopt the most common definition named after Riemann and Liouville, which is the natural generalization of the Cauchy formula for the n-fold primitive of the function $f(t)$. The Riemann–Liouville fractional derivative is defined as

$$D^{\alpha}f(t)=\begin{cases}\frac{\mathrm{d}^n f}{\mathrm{d}t^n}, & \alpha=n,\\ \frac{\mathrm{d}^n}{\mathrm{d}t^n}I^{n-\alpha}f(t), & 0\leq n-1<\alpha<n,\end{cases}\tag{22}$$

where $n\in\mathbb{N}$, $I^{\mu}f(t)$ is the Riemann–Liouville fractional integral of order μ, namely,

$$I^{\mu}f(t)=\frac{1}{\Gamma(\mu)}\int_0^t(t-s)^{\mu-1}f(s)\mathrm{d}s,\quad \mu>0,$$

$$I^0 f(t)=f(t)$$

and $\Gamma(z)$ is the standard Gamma function.

Similarly, the fractional partial derivative of the order α for the function $u(t,x)$ with respect to the variable t can be defined as follows.

Definition 3.1. *The Riemann–Liouville fractional partial derivative is defined by*

$$\partial_t^{\alpha}u(t,x)=\begin{cases}\frac{\partial^n u}{\partial t^n}, & \alpha=n,\\ \frac{1}{\Gamma(n-\alpha)}\frac{\partial^n}{\partial t^n}\int_0^t(t-s)^{n-\alpha-1}u(s,x)\mathrm{d}s, & 0\leq n-1<\alpha<n,\end{cases}\tag{23}$$

if the integral exists, where ∂_t^n is the usual partial derivative of integer order n.

There are also alternative definitions of the fractional derivative suggested by Caputo, Weyl and Riesz [10, 25, 33, 41, 45]. Each definition has its advantages and disadvantages. For instance, the Riemann–Liouville derivative of a constant is not zero. The Caputo derivative of a constant is zero, but it is defined only for differentiable functions. Note that some non-differentiable functions can have fractional derivatives of all the fractional orders smaller than one in the Riemann–Liouville sense [50].

We now consider the symmetry analysis for the FDE of the form

$$\partial_t^\alpha u = F(t, x, u, u_x, u_{xx}, u_{xxx}), \qquad 0 < \alpha < 1. \tag{24}$$

Given the one-parameter Lie group of infinitesimal transformations

$$t^* = t^*(t, x, u; \epsilon), \quad x^* = x^*(t, x, u; \epsilon), \quad u^* = u^*(t, x, u; \epsilon) \tag{25}$$

with the group parameter ϵ, the associated Lie algebra of (25) is spanned by vector fields

$$X = \tau\partial_t + \xi\partial_x + \eta\partial_u, \tag{26}$$

where

$$\tau = \left.\frac{\mathrm{d}t^*}{\mathrm{d}\epsilon}\right|_{\epsilon=0}, \quad \xi = \left.\frac{\mathrm{d}x^*}{\mathrm{d}\epsilon}\right|_{\epsilon=0}, \quad \eta = \left.\frac{\mathrm{d}u^*}{\mathrm{d}\epsilon}\right|_{\epsilon=0}.$$

According to the infinitesimal invariance criterion, Eq. (24) admits transformation group (25) (or equivalently, vector fields (26)) iff the prolonged vector field $\mathrm{pr}^{(\alpha,3)}X$ annihilates (24) on its solution manifold, namely,

$$\mathrm{pr}^{(\alpha,3)}X(\Delta)|_{\Delta=0} = 0, \qquad \Delta = \partial_t^\alpha u - F. \tag{27}$$

The operator $\mathrm{pr}^{(\alpha,3)}X$ takes the form (note that here we only keep the essential terms)

$$\mathrm{pr}^{(\alpha,3)}X = X + \eta^{\alpha,t}\partial_{\partial_t^\alpha u} + \eta^x\partial_{u_x} + \eta^{xx}\partial_{u_{xx}} + \eta^{xxx}\partial_{u_{xxx}}, \tag{28}$$

where

$$\begin{aligned}
\eta^{\alpha,t} &= D_t^\alpha(\eta) + \xi D_t^\alpha(u_x) - D_t^\alpha(\xi u_x) + D_t^\alpha(uD_t(\tau)) - D_t^{\alpha+1}(\tau u) + \tau D_t^{\alpha+1}(u),\\
\eta^x &= D_x(\eta) - u_t D_x(\tau) - u_x D_x(\xi),\\
\eta^{xx} &= D_x(\eta^x) - u_{xt} D_x(\tau) - u_{xx} D_x(\xi),\\
\eta^{xxx} &= D_x(\eta^{xx}) - u_{xxt} D_x(\tau) - u_{xxx} D_x(\xi),
\end{aligned}$$

the symbols D_t, D_x stand for the total derivatives with respect to t and x, respectively, and D_t^α is the total fractional derivative with respect to t.

In order to determine the vector fields (26) admitted by Eq. (24), it is necessary to obtain the concrete expression for $\mathrm{pr}^{(\alpha,3)}X$. Since the explicit expression for η^x, η^{xx} and η^{xxx} in (28) can be easily obtained in a standard fashion [42, 44], we concentrate on the expression for $\eta^{\alpha,t}$.

Using the generalized Leibnitz rule [33, 45] given by

$$D_t^\alpha(f(t)g(t)) = \sum_{n=0}^{\infty} \binom{\alpha}{n} D_t^n f(t) D_t^{\alpha-n} g(t), \tag{29}$$

where

$$\binom{\alpha}{n} = \frac{\Gamma(\alpha+1)}{\Gamma(n+1)\Gamma(\alpha+1-n)},\ D_t^0 f(t) = f(t),\ D_t^{n+1} f(t) = D_t(D_t^n f(t)),$$

we arrive at

$$\xi D_t^\alpha(u_x) - D_t^\alpha(\xi u_x) = -\sum_{n=1}^{\infty} \binom{\alpha}{n} D_t^n(\xi) D_t^{\alpha-n}(u_x),$$

and

$$\begin{aligned} D_t^\alpha(uD_t(\tau)) - D_t^{\alpha+1}(\tau u) + \tau D_t^{\alpha+1}(u) = &-\alpha D_t(\tau) D_t^\alpha u \\ &-\sum_{n=1}^{\infty} \binom{\alpha}{n+1} D_t^{n+1}(\tau) D_t^{\alpha-n}(u). \end{aligned}$$

Consequently, the relation

$$\begin{aligned} \eta^{\alpha,t} &= D_t^\alpha(\eta) - \alpha D_t(\tau) D_t^\alpha u - \sum_{n=1}^{\infty} \binom{\alpha}{n} D_t^n(\xi) D_t^{\alpha-n}(u_x) \\ &-\sum_{n=1}^{\infty} \binom{\alpha}{n+1} D_t^{n+1}(\tau) D_t^{\alpha-n}(u) \end{aligned}$$

is obtained. Now the problem of finding the expression for $\eta^{\alpha,t}$ reduces to determining the first summand $D_t^\alpha(\eta)$ of it. In view of the generalized chain rule for a composite function [41] of the form

$$\frac{\mathrm{d}^\alpha f(g(t))}{\mathrm{d}t^\alpha} = \sum_{n=0}^{\infty} \frac{U_n}{n!} \frac{\mathrm{d}^n f(z)}{\mathrm{d}z^n}\Big|_{z=g(t)},$$

where

$$U_n = \sum_{k=0}^{n} (-1)^k \binom{n}{k} g^k(t) \partial_t^\alpha (g^{n-k}(t)),$$

and the generalized Leibnitz rule (29), we have

$$D_t^\alpha(\eta) = \partial_t^\alpha \eta + \eta_u \partial_t^\alpha u - u \partial_t^\alpha \eta_u + \sum_{n=1}^{\infty} \binom{\alpha}{n} \partial_t^n \eta_u \partial_t^{\alpha-n} u + \mu \tag{30}$$

with

$$\mu = \sum_{n=2}^{\infty} \sum_{m=2}^{n} \sum_{k=2}^{m} \binom{\alpha}{n} \binom{n}{m} \frac{t^{n-\alpha} U_k}{k! \Gamma(n+1-\alpha)} \frac{\partial^{n-m+k} \eta}{\partial t^{n-m} \partial u^k}.$$

Note that $\mu = 0$ holds when the infinitesimal η is linear of the variable u, because of the existence of the derivatives $\partial^k \eta / \partial u^k$, $k \geq 2$ in the above expression. Based on above analysis, $\eta^{\alpha,t}$ takes the explicit form of

$$\begin{aligned} \eta^{\alpha,t} = {} & \sum_{n=1}^{\infty} \left[\binom{\alpha}{n} \partial_t^n \eta_u - \binom{\alpha}{n+1} D_t^{n+1}(\tau) \right] \partial_t^{\alpha-n} u - \sum_{n=1}^{\infty} \binom{\alpha}{n} D_t^n(\xi) \partial_t^{\alpha-n}(u_x) \\ & + \partial_t^\alpha \eta + (\eta_u - \alpha D_t(\tau)) \partial_t^\alpha u - u \partial_t^\alpha \eta_u + \mu. \end{aligned}$$

To determine the Lie symmetries (26) admitted by FDE (24), we act by the differential operator $\mathrm{pr}^{(\alpha,3)} X$ on Eq. (24) according to the invariance criterion (27), split the obtained relation by independent variables, equate these coefficients to zero, and solve the obtained over-determined system of linear PDEs and FDEs.

3.2. Lie Symmetries and Optimal System for the Fractional HD Equation

Turn now to derivation of Lie symmetries of Eq. (4) and construction of their optimal systems.

The invariance criterion for the fractional HD equation (4) reads

$$\left[\eta^{\alpha,t} - 3u^2 u_{xxx} \eta - u^3 \eta^{xxx} \right] |_{\partial_t^\alpha u = u^3 u_{xxx}} = 0.$$

By equating to zero the coefficients of linearly independent derivatives u_t, u_x, $u_{xx}, \cdots$, and $\partial_t^{\alpha-n} u_x$, $\partial_t^{\alpha-n} u$, we obtain the over-determined system of linear

PDEs and FDEs, which is called the determining equations:

$$\begin{aligned}
&\binom{\alpha}{n}\partial_t^n\eta_u - \binom{\alpha}{n+1}D_t^{n+1}(\tau) = 0, \qquad n = 1, 2, 3, \cdots,\\
&\tau_x = \tau_u = \xi_t = \xi_u = \eta_{uu} = 0,\\
&3u\xi_x - \alpha\tau_t u - 3\eta = 0,\\
&\partial_t^\alpha\eta - u\partial_t^\alpha\eta_u - u^3\eta_{xxx} = 0.
\end{aligned}$$

The determining equations above are easily integrated to yield the following general solution:

$$\tau = c_3 t + c_5,\ \xi = c_4 x^2 + c_2 x + c_1,\ \eta = (2c_4 x + c_2 - \frac{c_3\alpha}{3})u,$$

where c_i $(i = 1, 2, \cdots, 5)$ are arbitrary constants. Further, imposing the requirement that transformation (25) preserves the structure of the Riemann–Liouville fractional derivative operator, we have

$$\tau(t, x, u)|_{t=0} = 0.$$

This relation follows from the fact that the lower limit of the integral in (23) is fixed, and therefore the equation $t = 0$ should be invariant with respect to the transformation in question. Thus the following assertion holds.

Theorem 3.1. *The symmetry group of the fractional HD equation is generated by the vector fields*

$$X_1 = \partial_x, \quad X_2 = x\partial_x + u\partial_u, \quad X_3 = t\partial_t - \frac{\alpha}{3}u\partial_u, \quad X_4 = x^2\partial_x + 2xu\partial_u. \tag{31}$$

Note that X_1 is the space translation generator, X_2 and X_3 generate scaling symmetries, and X_4 is the conformal symmetry generator. The nonzero commutation relations of the algebra (31) read as

$$[X_1, X_2] = X_1,\ [X_1, X_4] = 2X_2,\ [X_2, X_4] = X_4.$$

It is also worth noting that the standard HD equation admits the five dimensional Lie symmetry group [18]

$$\langle\partial_x, x\partial_x + u\partial_u, t\partial_t - \frac{u}{3}\partial_u, x^2\partial_x + 2xu\partial_u, \partial_t\rangle.$$

The transformation group corresponding to the infinitesimal generator $x^2\partial_x + 2xu\partial_u$ is also given in [18].

We begin the procedure of similarity reduction of (4) by constructing the optimal system subalgebras of (31). Any transformation from the full symmetry group transforms a solution into another one, so we only need invariant solutions which are not related by symmetry group transformations. There are infinitely many subgroups of the symmetry group of a given DE, because any linear combination of infinitesimal generators is also an infinitesimal generator. In order to obtain the complete description of invariant solutions for a DE, we need to construct the subgroups which give essentially different solutions. In other words, we need to obtain the optimal system [22, 42, 44]. Constructing the optimal system of subgroups of the Lie group is evidently equivalent to finding an optimal system of subalgebras of the corresponding Lie algebra. For one dimensional subalgebras, the construction of optimal system reduces to the classification of the orbits of the adjoint representation. The simplest approach to this problem is taking the general element of the Lie algebra and simplifying it by a properly chosen adjoint transformation [44].

The action of the adjoint operator is given by the Lie series of the form

$$\mathrm{Ad}(\exp(\epsilon X_i))X_j = X_j - \epsilon[X_i, X_j] + \frac{\epsilon^2}{2}[X_i,[X_i,X_j]] - \cdots,$$

where $[X_i, X_j]$ is the standard commutator and ϵ is a parameter. In Table 3.1, we list the adjoint actions of the generators of algebra (31), where the (i,j)-th entry indicates $\mathrm{Ad}(\exp(\epsilon X_i))X_j$.

Table 1. The adjoint representation of (31)

$\mathrm{Ad}(\epsilon\cdot)$	X_1	X_2	X_3	X_4
X_1	X_1	$X_2 - \epsilon X_1$	X_3	$X_4 - 2\epsilon X_2 + \epsilon^2 X_1$
X_2	$\mathrm{e}^{\epsilon} X_1$	X_2	X_3	$\mathrm{e}^{-\epsilon} X_4$
X_3	X_1	X_2	X_3	X_4
X_4	$X_1 + 2\epsilon X_2 + \epsilon^2 X_4$	$X_2 + \epsilon X_4$	X_3	X_4

The one-dimensional optimal system of subalgebras of the algebra (31) are listed in the assertion below.

Theorem 3.2. *The one-dimensional optimal system of* (31) *is given by*

$$\begin{aligned} &V_1 = X_1, \quad V_2 = X_2, \quad V_3 = X_3, \quad V_4 = X_1 + X_3, \quad V_5 = X_1 - X_3, \\ &V_6 = X_1 + X_4, \quad V_7 = X_2 + \beta X_3 \ (\beta \neq 0), \quad V_8 = X_1 + X_4 + \gamma X_3 \ (\gamma \neq 0). \end{aligned} \tag{32}$$

Proof. Take V in the most general form $V = \sum_{i=1}^{4} a_i X_i$ and simplify it by suitable adjoint maps. Applying $\mathrm{Ad}(\exp(\epsilon_1 X_1))$ and $\mathrm{Ad}(\exp(\epsilon_2 X_4))$ to V yields

$$\tilde{V} = \sum_{i=1}^{4} \tilde{a}_i X_i = \mathrm{Ad}(\exp(\epsilon_2 X_4)) \circ \mathrm{Ad}(\exp(\epsilon_1 X_1))V,$$

where

$$\begin{aligned} \tilde{a}_1 &= a_1 - \epsilon_1 a_2 + \epsilon_1^2 a_4, \\ \tilde{a}_2 &= a_3 - 2\epsilon_1 a_4 + 2\epsilon_2(a_1 - \epsilon_1 a_2 + \epsilon_1^2 a_4), \\ \tilde{a}_4 &= a_4 + \epsilon_2(a_2 - 2\epsilon_1 a_4) + \epsilon_2^2(a_1 - \epsilon_1 a_2 + \epsilon_1^2 a_4). \end{aligned}$$

There are three possible inequivalent cases for the value of $a_2^2 - 4a_1a_4$. Consider these three cases separately.

Case 1. If $a_2^2 - 4a_1a_4 > 0$, then choosing the real root of the quadratic equation $a_4\epsilon_1^2 - a_2\epsilon_1 + a_1 = 0$ as ϵ_1 and putting $\epsilon_2 = a_4/(2\epsilon_1 a_4 - a_2)$, we obtain $\tilde{a}_1 = \tilde{a}_4 = 0$ and $\tilde{a}_2 \neq 0$. Consequently, V is equivalent to a multiple of

$$\tilde{V} = X_2 + \tilde{a}_3 X_3$$

and thus we have two vector fields X_2 and $X_2 + \beta X_3$ $(\beta \neq 0)$.

Case 2. Assuming $a_2^2 - 4a_1a_4 < 0$, we set $\epsilon_1 = a_3/(2a_4)$ and $\epsilon_2 = 0$. Hence $\tilde{a}_2 = 0$. Acting on this V by a suitable group generated by X_2, we can make the coefficients of X_1 and X_4 equal. So V is equivalent to a scalar multiple of

$$\tilde{V} = X_1 + X_4 + \tilde{a}_3 X_3.$$

Therefore V is equivalent to $X_1 + X_4$ or $X_1 + X_4 + \gamma X_3$ $(\gamma \neq 0)$.

Case 3. Given $a_2^2 - 4a_1a_4 = 0$, there are two subcases to consider.

If $a_1^2 + a_2^2 + a_4^2 \neq 0$, we can choose ϵ_1 and ϵ_2 so that $\tilde{a}_2 = \tilde{a}_4 = 0$ and what is more $\tilde{a}_1 \neq 0$. Thus V is equivalent to

$$\tilde{V} = X_1 + \tilde{a}_3 X_3.$$

Provided $\tilde{a}_3 \neq 0$, we use the group generated by X_2 to scale the coefficients of X_1 and X_3 and get that $\tilde{V}$ is equivalent to $X_1 + X_3$ or $X_1 - X_3$. If $\tilde{a}_3 = 0$, $\tilde{V}$ reduces to X_1.

Given $a_1 = a_2 = a_4 = 0$, we must have $a_3 \neq 0$ (otherwise, V is trivial). Now the only remaining vectors are the multiples of X_3 and the adjoint representation acts trivially on it. Then V is equivalent to X_3.

We have shown that any one-dimensional subspace of (31) is equivalent to one of the subspaces spanned by $V_1, \cdots, V_8$ in Theorem 3.2. To complete the proof of Theorem 3.2, we need to establish the mutual inequivalence of any two algebras in (32). Following [11], we introduce the adjoint invariant, which is a real valued function ϕ on a Lie algebra $\mathfrak{g}$ satisfying $\phi(\mathrm{Ad}(\mathfrak{g})X) = \phi(X)$ for all $X \in \mathfrak{g}$. If vectors X and Y generate conjugate one-dimensional subalgebras, the relation $\phi(X) = \phi(Y)$ holds for any invariant ϕ. Given the general vector $V = \sum_{i=1}^{4} a_i X_i$ of (31), ϕ can be expressed as a function of the coefficients $a_1, \cdots, a_4$.

Lemma 3.1. *$Q = a_2^2 - 4a_1a_4$ is an invariant.*

Proof. It is a common knowledge that the Killing form is invariant under the adjoint action. Through straightforward calculation, we have the killing form

$$K(V, V) = 2(a_2^2 - 4a_1a_4)$$

of the Lie algebra (31). Consequently, $Q = K(V, V)/2$ is an invariant. □

We would like to point out that the invariants of the full adjoint action are often used to simplify the general vector field when constructing the optimal system. Actually, the derivation of (32) is based on the fact that Q is an invariant.

Lemma 3.2. *$B = a_3$ and*

$$C = \begin{cases} sign\, a_1, & a_2 = a_4 = 0, \\ 0, & otherwise \end{cases}$$

are invariants.

Proof. This lemma follows directly from the Table 3.1. □

Now we determine the invariants Q, B and C for each algebra V_i $(i = 1, 2, \cdots, 8)$ in (32) and give the results in Table 3.2.

Table 2. The invariants of the algebra (32)

	V_1	V_2	V_3	V_4	V_5	V_6	V_7	V_8
Q	0	1	0	0	0	-4	1	-4
B	0	0	1	1	-1	0	β	γ
C	1	0	0	1	1	0	0	0

From this table, we can easily verify the inequivalence of V_i. This completes the proof of the Theorem 3.2.

3.3. Similarity Reductions of the Fractional HD Equation

Below we give the definition of a group invariant solution of FDE, which is similar to that of PDE.

Definition 3.2. $u = \theta(t, x)$ *is an invariant solution of Eq.* (24) *corresponding to the infinitesimal generator* (26) *iff*

- $u = \theta(t, x)$ *satisfies Eq.* (24)*;*
- $u = \theta(t, x)$ *is an invariant surface of* (26)*, namely, it obeys the invariant surface condition*

$$\tau(t, x, \theta)\theta_t + \xi(t, x, \theta)\theta_x = \eta(t, x, \theta).$$

In what follows, we perform similarity reductions, derive reduced nonlinear fractional ODEs, and classify the corresponding group-invariant solutions of the fractional HD equation (4) for each inequivalent subalgebra listed in (32).

Case 1. $V_1 = \partial_x$. Integration of the invariant surface condition

$$\frac{dt}{0} = \frac{dx}{1} = \frac{du}{0}$$

gives the similarity variables $t,\ u$. Thus we have the ansatz $u = g(t)$. Inserting it into Eq. (4) yields the reduced fractional ODE

$$\partial_t^\alpha g(t) = 0, \tag{33}$$

which implies that $g(t) = a_1 t^{\alpha-1}$. Here and hereafter a_i are arbitrary real constants. Consequently, Eq. (4) has group-invariant solutions of the form

$$u = a_1 t^{\alpha-1}.$$

Case 2. $V_2 = x\partial_x + u\partial_u$. Solving the invariant surface condition, we get two independent similarity variables t and u/x. Substituting the ansatz $u = xg(t)$ into Eq. (4) yields reduced equation (33). Integrating the latter, we get the group-invariant solution

$$u = a_2 x t^{\alpha-1}.$$

of Eq. (4) corresponding to V_2.

Case 3. $V_3 = t\partial_t - \frac{\alpha}{3}u\partial_u$. Solving the invariant surface condition

$$\frac{\mathrm{d}t}{t} = \frac{\mathrm{d}x}{0} = \frac{\mathrm{d}u}{-\frac{\alpha}{3}u}$$

yields the similarity variables x and $t^{\frac{\alpha}{3}}u$. Inserting the ansatz $u = t^{-\frac{\alpha}{3}}g(x)$ into Eq. (4), we arrive at the reduced equation

$$g^2(x)g'''(x) = \frac{\Gamma(1-\frac{\alpha}{3})}{\Gamma(1-\frac{4}{3}\alpha)}. \tag{34}$$

Consequently, Eq. (4) has group invariant solution of the form $u = t^{-\frac{\alpha}{3}}g(x)$, where $g(x)$ satisfies Eq. (34).

Case 4. $V_4 = t\partial_t + \partial_x - \frac{\alpha}{3}u\partial_u$. In this case, the similarity variables are $\mathrm{e}^{-x}t$ and $\mathrm{e}^{\frac{\alpha}{3}x}u$. In view of the reduction ansatz $u = \mathrm{e}^{-\frac{\alpha}{3}x}g(\mathrm{e}^{-x}t)$, we arrive at

$$\partial_z^\alpha g(z) = -g^3(z)\Big[\frac{\alpha^3}{27}g(z) + (\frac{\alpha^2}{3} + \alpha + 1)zg'(z) + (\alpha+3)z^2g''(z) + z^3g'''(z)\Big],$$

where $z = \mathrm{e}^{-x}t$.

Case 5. $V_5 = -t\partial_t + \partial_x + \frac{\alpha}{3}u\partial_u$. Integrating the invariant surface condition

$$\frac{\mathrm{d}t}{-t} = \frac{\mathrm{d}x}{1} = \frac{\mathrm{d}u}{\frac{\alpha}{3}u},$$

the similarity variables $\mathrm{e}^x t,\ \mathrm{e}^{-\frac{\alpha}{3}x}u$ are obtained. The group invariant solution has the form

$$u = \mathrm{e}^{\frac{\alpha}{3}x} g(\mathrm{e}^x t),$$

where $g(z)$ satisfies the equation

$$\partial_z^\alpha g(z) = g^3(z)\Big[\frac{\alpha^3}{27}g(z) + (\frac{\alpha^2}{3} + \alpha + 1)zg'(z) + (\alpha+3)z^2g''(z) + z^3g'''(z)\Big].$$

Case 6. $V_6 = (1+x^2)\partial_x + 2xu\partial_u$. Considering the invariant surface condition

$$\frac{\mathrm{d}t}{0} = \frac{\mathrm{d}x}{1+x^2} = \frac{\mathrm{d}u}{2xu},$$

we arrive at the similarity variables t and $u/(1+x^2)$. Inserting the ansatz $u = (1+x^2)g(t)$ into Eq. (4) yields Eq. (33). Hence we get the group-invariant solution

$$u = a_3(1+x^2)t^{\alpha-1}$$

of (4) that corresponds to V_6. Since the linear combination of the invariant solutions is still an invariant solution, we have the following exact solutions

$$u = (a_3x^2 + a_2x + a_1)t^{\alpha-1}$$

of Eq. (4).

Case 7. $V_7 = \beta t\partial_t + x\partial_x + (1 - \frac{\alpha\beta}{3})u\partial_u$. The similarity variables for this generator are $x^{-\beta}t$ and $x^{\frac{\alpha\beta}{3}-1}u$. We now look for a similarity reduction for (4) in consideration of the reduction ansatz

$$u = x^{1-\frac{\alpha\beta}{3}} g(x^{-\beta}t).$$

Substituting the above expression into Eq. (4) leads to the reduced equation

$$\partial_z^\alpha g(z) = -g^3(z)\Big[\frac{\alpha\beta(\alpha^2\beta^2-9)}{27}g(z) + \frac{\beta(\alpha^2\beta^2+3\beta^2\alpha+3\beta^2-3)}{3}zg'(z)$$
$$+ (\alpha+3)\beta^3z^2g''(z) + \beta^3z^3g'''(z)\Big],$$

where $z = x^{-\beta}t$.

Case 8. $V_8 = \gamma t\partial_t + (1+x^2)\partial_x + (2x - \frac{\alpha\gamma}{3})u\partial_u$. Due to the invariant surface condition

$$\frac{\mathrm{d}t}{\gamma t} = \frac{\mathrm{d}x}{1+x^2} = \frac{\mathrm{d}u}{(2x - \frac{\alpha\gamma}{3})u},$$

we obtain the following similarity variables

$$\mathrm{e}^{-\gamma \arctan x}t, \qquad \frac{\mathrm{e}^{\frac{\alpha\gamma}{3}\arctan x}}{1+x^2}u.$$

Thus Eq. (4) has group invariant solution of the form

$$u = (1+x^2)\mathrm{e}^{-\frac{\alpha\gamma}{3}\arctan x}g(\mathrm{e}^{-\gamma\arctan x}t),$$

where $g(z)$ satisfies the equation

$$\partial_z^\alpha g(z) = -g^3(z)\Big[\frac{\alpha\gamma(\alpha^2\gamma^2+36)}{27}g(z) + \frac{\gamma(\alpha^2\gamma^2+3\alpha\gamma^2+3\gamma^2+12)}{3}zg'(z) + (\alpha+3)\gamma^3z^2g''(z) + \gamma^3z^3g'''(z)\Big].$$

Conclusion

In this chapter we presented a review of our recent results on group analysis of nonlinear evolution equations in one spatial variable.

In Sec. 2, we generalize the Lie group classification method of [59] and make it applicable to contact symmetries. As a result, we develop an efficient algebraic approach to classification of contact symmetry groups of nonlinear evolution equations. It enables obtaining exhaustive description of all second-order evolution equations that possess contact symmetry algebras containing semi-simple subalgebras.

The number of inequivalent realizations obtained here is lower than that of the realizations based on point symmetries [4, 60]. The reason for this is that contact transformations provide stronger equivalence relation, i.e., some realizations, which are inequivalent with respect to point transformations, may be equivalent with respect to contact ones [43]. Take semi-simple algebras, for instance. There are five realizations of the $\mathfrak{sl}(2,\mathbb{R})$ algebra in [60]. Among them, two realizations in Theorem 3.1 are equivalent to $\mathfrak{sl}^3(2,\mathbb{R})$, and three realizations from Lemma 4.2 can be transformed to $\mathfrak{sl}^1(2,\mathbb{R})$ or $\mathfrak{sl}^2(2,\mathbb{R})$. More specifically, the first realization, $\langle \partial_t, 2t\partial_t + x\partial_x, -t^2\partial_t - tx\partial_x + x^2\partial_u\rangle$, in Theorem 3.1 of [60], can be mapped into $\langle u_t, -2tu_t+2xu_1, t^2u_t-2txu_1-u_1\rangle$ by the contact transformation

$$\tilde{t}=t, \quad \tilde{x}=u_1^{\frac{1}{2}}, \quad \tilde{u}=-u+xu_1.$$

Evidently, any algebra containing semi-simple subalgebra in [4, 60] can be reduced to one of our realizations by a suitable contact transformation.

It is important to emphasize that our approach can be efficiently applied to hyperbolic PDEs as well. We have classified the broad classes of nonlinear wave equations admitting nontrivial point symmetries [26]. We intend to apply the approach described in this chapter to classify nonlinear hyperbolic PDEs admitting nontrivial contact symmetries.

Note that the relation (8) holds for evolution equations of an arbitrary order [31] and our approach can be applied to higher dimensional evolution equations as well.

We intend to utilize the obtained results to study potential symmetries of the class (1), based on the relation between contact and potential symmetries [19, 46, 47, 58]. We prove in [19] the following assertion:

Theorem 4.1. *Let L be the algebra of contact symmetries of evolution Eq.* (1) *such that, (i) dim(L) ≥ 2, (ii) $[L,L]\neq 0$. Then Eq.* (1) *can be mapped to another evolution equation, belonging to the class* (1)*, which admits potential symmetries.*

Since any semi-simple algebra is non-commutative, every realization obtained here will lead to a potential symmetry. To construct all possible second-order evolution PDEs possessing potential symmetries, one needs to classify all inequivalent non-commutative subalgebras of the symmetry algebras constructed above and then follow the procedure suggested in [19, 58] to obtain a wide range of Eqs. (1) admitting potential symmetries.

In Sec. 3, we demonstrate the efficiency of the classical Lie group approach to analysis of FDEs. We perform exhaustive analysis of subgroup structure of the invariance group admitted by the time-fractional Harry-Dym equation with the Riemann–Liouville derivative. As an application of the so obtained classification of subgroups, we construct all inequivalent reduced fractional ODEs and also obtain some explicit exact solutions of the equation in consideration.

Integration of reduced fractional ODEs, itself, is a very challenging problem. In order to obtain the complete set of group-invariant solutions of Eq. (4), one needs to develop a symmetry or numeric approach to solving reduced fractional ODEs.

References

[1] R. L. Anderson and N. H. Ibragimov, *Lie-Bäcklund Transformations in Applications* (SIAM, Philadelphia, PA, 1979).

[2] R. Bagley and P. Torvik, "A theoretical basis for the application of fractional calculus to viscoelasticity", *J. Rheol.* 27, 201–210 (1983).

[3] A. O. Barut and R. Rączka, *Theory of Group Representations and Applications* (World Scientific, Singapore, 1986).

[4] P. Basarab-Horwath, V. Lahno and R. Zhdanov, "The structure of Lie algebras and the classification problem for partial differential equations", *Acta Appl. Math.* 69, 43–94 (2001).

[5] A. H. Bhrawy, T. M. Taha and J. A. T. Machado, "A review of operational matrices and spectral techniques for fractional calculus". *Nonlinear Dyn.* 81, 1023–1052 (2015).

[6] G. W. Bluman and S. Kumei, *Symmetries and Differential Equations* (Springer-Verlag, New York, 1989).

[7] G. W. Bluman, A. F. Cheviakov and S. C. Anco, *Applications of Symmetry Methods to Partial Differential Equations* (Springer, New York, 2010).

[8] E. Buckwar and Yu. Luchko, "Invariance of a partial differential equation of fractional order under the Lie group of scaling transformations", *J. Math. Anal. Appl.* 227, 81–97 (1998).

[9] J. E. Campbell, *Introductory Treatise on Lie's Theory of Finite Continuous Transformation Groups* (The Clarendon Press, Oxford, 1903).

[10] A. Carpinteri and F. Mainardi, *Fractals and Fractional Calculus in Continuum Mechanics* (Springer, Wien, 1997).

[11] K. S. Chou and C. Z. Qu, "Optimal systems and group classification of (1+2)-dimensional heat equation", *Acta Appl. Math.* 83, 257–287 (2004).

[12] V. D. Djordjevic and T. M. Atanackovic, "Similarity solutions to nonlinear heat conduction and Burgers Korteweg-de Vries fractional equations", *J. Comput. Appl. Math.* 212, 701–714 (2008).

[13] R. K. Gazizov, A. A. Kasatkin and S. Yu. Lukashchuk, "Continuous transformation groups of fractional differential equations", *Vestnik USATU* 9, 125–135 (2007).

[14] R. K. Gazizov, A. A. Kasatkin and S. Yu. Lukashchuk, "Symmetry properties of fractional diffusion equations", *Phys. Scr.* 136, 014016 (2007).

[15] R. K. Gazizov, N. H. Ibragimov and S. Yu. Lukashchuk, "Nonlinear self-adjointness, conservation laws and exact solutions of time-fractional Kompaneets equations", *Commun. Nonlinear Sci. Numer. Simulat.* 153–163, 014016 (2015).

[16] J. H. He, "Approximate analytical solution for seepage flow with frctional derivatives in porous media", *Comput. Methods Appl. Mech. Eng.* 167, 57–68 (1998).

[17] R. H. Heredero and P. J. Olver, "Classification of invariant wave equations", *J. Math. Phys.* 37, 6414–6438 (1996).

[18] W. Hereman, "SYMMGRP. MAX and other symbolic programs for lie symmetry analysis of partial differential equation", *Lectures in Appl. Math.* 29, 241–257(1993).

[19] Q. Huang, C. Z. Qu and R. Zhdanov,"Group-theoretical framework for potential symmetries of evolution equations", *J. Math. Phys.* 52, 023514 (2011).

[20] Q. Huang and S. F. Shen, "Lie symmetries and group classification of a class of time fractional evolution systems", *J. Math. Phys.* 56, 123504 (2015).

[21] P. E. Hydon, "How to find discrete contact symmetries", *J. Non. Math. Phys.* 5, 405–416 (1998).

[22] N. H. Ibragimov, *Transformation Groups Applied to Mathematical Physics* (D. Reidel, Dordrecht, 1985).

[23] N. H. Ibragimov and S. V. Khabirov, "Contact transformation group classification of nonlinear wave equations", *Nonlinear Dynam.* 22, 61–71 (2000).

[24] A. H. Kara and F. M. Mahomed, "Relationship between symmetries and conservation laws", *Int. J. Theor. Phys.* 39, 23–40 (2000).

[25] A. Kilbas, H. M. Srivastava and J. J. Trujillo, *Theory and Applications of Fractional Differential Equations* (Elsevier Science B.V, Amsterdam , 2006).

[26] V. Lahno, R. Zhdanov and O. Magda, "Group classification and exact solutions of nonlinear wave equations", *Acta Appl. Math.*, 91, 253–313 (2006).

[27] S. Lie, Ph. D. thesis, University of Christiana, 1871.

[28] S. Lie, "*Ü*ber die Integration durch bestimmte Integrable von einer Klasse linear partialler Differentialgleichungen", *Arch. Math.* 6, 328 (1881).

[29] H. Z. Liu, "Complete group classifications and symmetry reductions of the fractional fifth-order KdV types of equations", *Stud. Appl. Math.* 131, 317–330 (2013).

[30] A. Yu. Loskutov and A. S. Mikhalov, *Introduction to Synergetics* (Nauka, Moscow, 1990).

[31] B. A. Magadeev, "On group classification of nonlinear evolution equations", *Algebra i Analiz* 5, 141–156 (1993).

[32] M. M. Meerschaert, H. P. Scheffler and C. Tadjeran, "Finite difference methods for two-dimensional fractional dispersion equation", *J. Comput. Phys.* 211, 249–261 (2006).

[33] K. S. Miller and B. Ross, *An Introduction to the Fractional Calculus and Fractional Differential Equations* (Wiley, New York, 1993).

[34] S. Momani and Z. Odibat, "Analytical solution of a time fractional Navier-Stokes equation by Adomian decompostion method", *Appl. Mech. Comput.* 177, 488–494 (2006).

[35] S. Momani and Z. Odibat, "Homotopy perturbation method for nonlinear partial differential equations of fractional order", *Phys. Lett. A* 365, 345–350 (2007).

[36] E. Momoniat, "True contact transformations admitted by a class of non-linear wave equations", *Int. J. Non. Mech.* 36, 705–708 (2001).

[37] E. Momoniat, "An equation admitting infinite true contact transformations", *J. Math. Anal. Appl.* 282, 668–672 (2003).

[38] O. I. Morozov, "Contact-equivalence problem for linear hyperbolic equations", *J. Math. Sci.* 135, 2680–2694 (2006).

[39] A. C. Newell and J. A. Whitehead, "Finite bandwidth, finite amplitude convection", *J. Fluid Mech.* 38, 279–303 (1969).

[40] Z. Odibat and S. Momani, "A generalized differential transform method for linear partial differential equations of fractional order", *Appl. Math. Lett.* 21, 194–199 (2008).

[41] K. B. Oldham and F. Spanier, *The Fractional Calculus* (Academic Press, New York, 1974).

[42] P. J. Olver, *Application of Lie Groups to Differential Equations* (Springer-Verlag, New York, 1986).

[43] P. J. Olver, *Equivalence, Invariants, and Symmetry* (Cambridge University Press, Cambridge, 1995).

[44] L. V. Ovsiannikov, *Group Analysis of Differential Equations* (Academic, New York, 1982).

[45] I. Podlubny, *Fractional Differential Equations* (Academic Press, San Diego, CA, 1999).

[46] R. O. Popovych and A. M. Samoilenko, "Local conservation laws of second-order evolution equations", *J. Phys. A* 41, 362002 (2008).

[47] R. O. Popovych and A. Sergyeyev, "Conservation laws and normal forms of evolution equations", *Phys. Lett. A* 374, 2210–2217 (2010).

[48] E. Pucci and G. Saccomandi, "Contact symmetries and solutions by reduction of partial differential equations", *J. Phys. A: Math. Gen.* 27, 177–184 (1994).

[49] R. Sahadevan and T. Bakkyaraj, "Invariant analysis of time fractional generalized Burgers and Korteweg-de Vries equation", *J. Math. Anal. Appl.* 393, 341–347 (2012).

[50] S. G. Samko, A. A. Kilbas and O. I. Marichev, *Fractional Integrals and Derivatives* (Gordon and Breach, Yverdon, 1993).

[51] V. V. Sokolov, "On the structure of the symmetry algebra for one-field evolution equation", *Soviet Math. Dokl.* 35, 635–638 (1987).

[52] V. V. Sokolov, "On the symmetries of evolution equations", *Russian Math. Surveys* 43, 165–204 (1988).

[53] P. Turkowski, "Low-dimensional real Lie algebras", *J. Math. Phys.* 29, 2139–2144 (1988).

[54] G. W. Wang, X. Q. Liu and Y. Y. Zhang, "Lie symmetry analysis to the time fractional generalized fifth-order KdV equation", *Commun. Nonlinear Sci Numer Simulat* 18, 2321–2326, 2013.

[55] G. Wu and E. W. M. Lee, "Fractional variational iteration method and its application", *Phys. Lett. A* 374, 2506–2509 (2010).

[56] O. Yu. Yefimova and N. A. Kudryashov, "Exact solutions of the Burgers-Huxley equation", *J. Appl. Math. Mech.* 68, 413–420 (2004).

[57] G. M. Zaslavsky, *Hamiltonian Chaos and Fractional Dynamics* (Oxford Univ. Press, Oxford, 2005).

[58] R. Zhdanov, "On relation between potential and contact symmetries of evolution equations", *J. Math. Phys.* 50, 053522 (2009).

[59] R. Z. Zhdanov and V. I. Lahno, "Group classification of heat conductivity equations with a nonlinear source", *J. Phys. A: Math. Gen.* 32, 7405–7418 (1999).

[60] R. Zhdanov and V. Lahno, "Group classification of the general second-order evolution equation: semi-simple invariance groups", *J. Phys. A: Math. Theor.* 40, 5083–5103 (2007).

In: Emerging Concepts in Evolution Equations ISBN: 978-1-53610-861-3
Editor: Carolyn Murphy, pp. 69-93

Chapter 3

A METHOD FOR EXPONENTIALLY STABILIZING A CLASS OF 1-D PDES

Abdelhadi Elharf [*] ***and Ahmed Sani*** [†]
Department of Mathematics,
Regional Center of the Education Professions and Formation,
Agadir, Morocco
Department of Mathematics, Faculty of Sciences
Ibn Zohr University, Agadir, Morocco

Abstract

In this chapter, we address the problem of exponential stabilization of a class of 1-D PDEs with Dirichlet boundary control. We first reformulate the control problem in semigroups context on a suitable state space. We also convert the initial system to an auxiliary one by the use of a Volterra transformation with a kernel satisfying an appropriate PDE. Secondly, we prove the exponential stabilization of the target system and we show some useful spectral properties. Thirdly, we deal with the exponential stability of the initial system. In semigroup context, this corresponds to construct a boundary controller which causes an exponential decay of the semigroup $S(\cdot)$ representing the closed-loop system. Precisely, $\|S(t)\| = O(e^{-\lambda t}),\ t \to \infty$, for some $\lambda > 0$. The proposed method provides an explicit solution of the Lyapunov's equation by means of the kernel PDE

[*]E-mail address: a.elharfi@gmail.com.
[†]E-mail address: ahmedsani82@gmail.com.

and the parameters of the system in consideration. Further, the technics of constructing the feedback law offer operational information about the stabilization mechanism.

Keywords: feedback stabilization, generator of a C_0-semigroup

1. Formulation of the Control Problem

In this chapter we are concerned with dynamical systems governed by the following hyperbolic PDE

$$\begin{cases} \ddot{y}(t,x) = (\tau(x)y_x(t,x))_x(t,x) + p(x)y_x(t,x) + q(x)y(t,x), & \text{in } (0,\infty)\times(0,1), \\ M\ddot{y}(t,0) - \tau(0)y_x(t,0) = 0, \quad y(t,1) = u(t) & \text{in } (0,\infty), \\ y(0,x) = y^0(x), \quad \dot{y}(0,x) = y^1(x) & \text{in } (0,1), \end{cases} \tag{1}$$

with coefficients satisfying

$$\begin{array}{l} \bullet\ q \in C^1([0,1]), \quad \tau,\ p \in C^2([0,1]), \quad M,\ m > 0, \\ \bullet\ \tau \text{ is increasing with } \tau(0) > 0. \end{array} \tag{H}$$

Here u is a control placed at the extremity $x = 1$, the functions y^0, y^1 are the initial conditions.

The open-loop system, i.e. (1) with $u = 0$, is generally instable. In fact, depending on the parameters p, q, the perturbing term $py_x + qy$ generates eigenvalues of the open-loop system in the right half of the complex plan. This part of the spectrum influences the system stability, see e.g. [14, 15] or [20] for $\tau = 1$, $p = 0$, $q = const$ with $q > \frac{3\pi^2}{4}$. On the other hand, in spite of the simplicity of system (1), the *exponential* stability is often unfulfilled in open-loop systems and difficult to reach even by states feedback. In the literature, only the *strong* stability is achieved for such a system subject to different boundary controls [1, 4, 6, 9, 27]. The studies dealing with the exponential stability for subclasses of system (1) are rather rare [3, 5, 25]. These considerations motivate us to investigate the problem of exponential stabilizability of system (1). Precisely,

Problem 1.1. *Find a feedback law of the form* $y(t,1) = u(t)$ *which exponentially stabilizes system* (1).

Different mathematical tools have been applied in order to deal with the stabilizability of the control systems such as spectral methods [17, 26], the linear quadratic regulator approach [18], the multiplier technic [19], the microlocal analysis [4] and Lyapunov functionals [27]. All of the above mentioned tools may be reformulated in semigroups theory. Strongly continuous semigroups (C_0-semigroups for short) remain an efficient tool for investigating the stabilizability of open-loop systems [11]. They not only yield well-posedness results but also provide qualitative properties. In semigroups context, Problem 1.1 corresponds to the designation of a boundary feedback law which produces an exponential decay of the semigroup $\mathcal{S}$ representing the resulting closed-loop, i.e., there exists $\lambda > 0$ such that $\|\mathcal{S}(t)\| = O(e^{-\lambda t})$, $t \to \infty$. This last estimate turns out to move all of the eigenvalues of the open-loop system to the left complex half plan by the use of an appropriate feedback law. We should emphasize here that the strategy based on the following perturbation argument: *the stability property persists to small perturbations,* is not applicable in this situation. In fact, the perturbation term $q(.)y(.)$ is not is not small enough because no assumption on the bounds of $q(.)$ (i.e $\inf q(t)$ and $\sup q(t)$) is not required. As we will see, the method we propose allows to overcome this difficulty and provides an explicit resolution of the Lyapunov equation of the stabilizability. Furthermore, it offers information about the stabilization process. Note that the current chapter is mainly based on the recent paper Elharfi [10] in which the reader finds additional results and more details.

2. Preliminaries

2.1 Exponential Stability in Semigroups Context

The purpose of this section is to present a look on the exponential stability/stabilizability concepts in the framework of semigroups theory as well as to survey the basic existing results. We emphasize on the Lyapunov's stability criterion that we will use to prove the exponential stability of the proposed system. We suppose the reader is familiar with the basic concepts of semigroups theory.

Let A be the generator of a C_0-semigroup $T(\cdot)$ on a Banach space X, and consider the abstract Cauchy problem

$$\begin{cases} \dot{v}(t) = Av(t), \quad t \geq 0, \\ v(0) = x, \end{cases} \tag{ACP}$$

Recall that for a fixed $x \in X$ the solution of (ACP) is given by $v(t) = T(t)x$. Which corresponds to the orbit of (ACP) starting from x. Then, the asymptotic behavior of the solution $v(\cdot)$ of (ACP) is closely related to the behavior of the semigroup $T(\cdot)$ for large $t > 0$, i.e., the existence (or nonexistence) of

$$\lim_{t\to\infty} T(t)$$

where the limit will be understood in various topologies. It is convenient for the rest to first make precise what we mean by the exponential stability of a C_0-semigroup $T(\cdot)$.

Definition 2.1. *A C_0-semigroup $T(\cdot)$ on a Banach space X is called exponentially stable if there exists $\varepsilon > 0$ such that*

$$\lim_{t\to\infty} e^{\varepsilon t}\|T(t)\| = 0$$

Exponential stability of a C_0-semigroup means that the closure of $\mathcal{T} = \{T(t),\ t \geq 0\}$ for uniform topology is the set

$$\mathcal{T} \cup \{0\}.$$

We now survey characterizations of the exponentially stability of a C_0-semigroup on Banach and Hilbert spaces through spectral and resolvent properties of the generator. An obvious one is obtained by looking at the growth bound $\omega_0(T)$ of the semigroup. A C_0-semigroup $T(\cdot)$ is exponentially stable if and only if $\omega_0(T) < 0$.

It is clear that if $\omega_0(T) < 0$, then there exist constants $\varepsilon > 0$ and $M \geq 1$ such that $\|T(t)x\| \leq Me^{-\varepsilon t},\ \ t \geq 0$. Hence, for all $p \geq 1$

$$\int_0^\infty \|T(t)x\|^p dt < \infty, \quad \forall x \in X.$$

This implies that for each $x \in X$ the orbit map $T(\cdot)x$ belongs to $L^p(\mathbb{R}^+, X)$ for all $p \in [1, +\infty)$. Datko [7] states that also the converse implication holds for $p = 2$. This result was generalized by Pazy [22], who showed that the exponent $p = 2$ may be replaced by any $1 \leq p < \infty$. Recently, Littman [21] showed that $\omega_0(T) < 0$ if and only if there exists a continuous, increasing function $\phi : \mathbb{R}^+ \to \mathbb{R}^+$ with $\phi(0) = 0$ and $\phi(t) > 0$ for all $t > 0$, such that

$$\int_0^\infty \phi\left(\|T(t)x\|\right) dt < \infty, \quad \forall x \in X.$$

Obviously, the Pazy's result is recovered from the function $\phi(t) = t^p$.
A weak version of the Datko-Pazy theorem is obtained by Weiss [29] who showed that a C_0-semigroup $T(\cdot)$ on a Banach space X is exponentially stable if and only if for some $p \geq 1$.

$$\int_0^\infty |\langle T(t)x, x'\rangle|^p dt < \infty, \quad \text{for all } x \in X \text{ and } x' \in X',$$

These characterizations of the exponential stability, as nice as they are, require the explicit knowledge of the semigroup and its orbits. However, in practical situations only the generator (or its resolvent) is often given. Therefore, direct characterizations of the exponential stability of the semigroup $T(\cdot)$ in terms of its generator A are more desirable. The first result in this direction goes back to 1892 due to M.A. Lyapunov. He showed that the semigroup $T(\cdot) := e^{\cdot A}$ generated by a matrix$A \in M_n(\mathbb{C})$ is exponentially stable if and only if all eigenvalues of A have non positive real parts, i.e., $\Re(\lambda) < 0$ for all $\lambda \in \sigma(A)$, see e.g. [11, p.12].
This theorem is of great theoretical and practical importance since it gives a property of the semigroup, namely the exponential stability, in terms of a spectral property of its generator, that is the spectrum. The Lyapunov's necessary and sufficient condition for a matrix to be exponentially stable can be extended to exponential stability of semigroups of bounded linear operators. Hille and Phillips [16, Sec 2,3], discovered that an analogous statement does not hold for strongly continuous semigroups $T(\cdot)$ with unbounded generator A on Banach spaces. By looking at the Lyapunov's theorem in finite dimensional space, one hopes that $s(A) < 0$ yields a characterization of the exponential stability. However, such a condition is not sufficient in infinite dimensional setting, the reason is the failure of the equality,

$$\omega_0(T) = s(A). \tag{2}$$

Clearly, if (2) holds then $s(A) < 0$ is equivalent to the exponential stability of $T(\cdot)$. In this case, the negativity of the spectral bound yields a characterization of the behavior of the semigroup in terms of its generator and its spectrum $\sigma(A)$. But in general, the equality (2) does not hold. Precisely, we have

$$-\infty \leq s(A) \leq \omega_0(T).$$

By looking at the exponential stability definition, one observes that the topological structure of the space X is involved. So it is natural to check the exponential stability on particular Banach spaces, namely Hilbert spaces. One of interesting characterization of the exponential stability on Hilbert space in terms of the the resolvent of the generator is due to Gearhart [13]. A C_0-semigroup $T(\cdot)$ on a Hilbert space H with generator A is exponentially stable, if and only if, $\mathbb{C}_0 \subseteq \rho(A)$ and

$$M := \sup_{Re\lambda > 0} \|R(\lambda, A)\| < \infty. \tag{3}$$

(Or equivalently, $R(\cdot, A) \in H^{\infty}(\mathcal{L}(X))$) . We now give from Datko [7] a characterization of the exponential stability on Hilbert. It constitutes the basic mathematical tool for proving the exponential stability in the current study.

Theorem 2.2. *Let A be the generator of a C_0-semigroup $T(\cdot)$ on a Hilbert space H. Then, the following assertions are equivalent*
(a) $T(\cdot)$ is exponentially stable.
(b) There is a positive operator $P \in \mathcal{L}(X)$ such that

$$\langle Ax, Px\rangle + \langle Px, Ax\rangle = -\langle x, x\rangle, \qquad \text{for all } x \in D(A). \tag{4}$$

(c) There is a positive operator $P \in \mathcal{L}(X)$ such that

$$\langle Ax, Px\rangle + \langle Px, Ax\rangle \leq -\langle x, x\rangle \qquad \text{for all } x \in D(A). \tag{5}$$

Identity (4) *is known as Lyapunov's equation.*

Proof. (a) $\Rightarrow$ (b). Since $T(\cdot)$ is exponentially stable then, there exist $\alpha > 0$ and $M > 0$ such that

$$\|T(t)\| \leq Me^{-\alpha t}, \quad \text{for all } t \geq 0.$$

Then, the operator,

$$Px = \int_0^{\infty} T(t)'T(t)x dt, \qquad x \in X,$$

is well-defined form X into X. Moreover, for all $x,\ y \in X$,

$$\langle Px, y\rangle = \int_0^{\infty} \langle T(t)x, T(t)y\rangle dt \leq \frac{M^2}{2\alpha}\|x\|\|y\|$$

Hence, P is bounded. Further, if $Px = 0$ we infer from

$$\langle Px, x\rangle = \int_0^{\infty} \|T(t)x\|^2 dt,$$

that $T(t)x = 0$ a.e. on $[0, \infty)$. Which implies, by the strong continuity of $T(\cdot)x$, that

$$x = \lim_{t\to\infty} T(t)x = 0.$$

Consequently, $P > 0$. On the other hand, for all $x \in D(A)$ we have

$$\begin{aligned}\langle Px, Ax\rangle + \langle Ax, Px\rangle &= \int_0^\infty \langle T(t)x, T(t)Ax\rangle dt + \langle AT(t)x, T(t)x\rangle dt \\ &= \int_0^\infty \frac{d}{dt}\|T(t)x\|^2 dt = \|T(\infty)x\|^2 - \|T(0)x\|^2 = -\|x\|^2\end{aligned}$$

(b) $\Rightarrow$ (c) is obvious.
(c) $\Rightarrow$ (a). Introduce the following Lyapunov functional

$$F(t, x) = \langle PT(t)x, T(t)x\rangle$$

where P is the positive operator due to (c). Since $P \geq 0$ then $F(t, x) \geq 0$ for all $t \geq 0$. Differentiating F with respect to t, we obtain by the use of (c),

$$\begin{aligned}\frac{dF}{dt}(t, x) &= \langle PAT(t)x, T(t)x\rangle + \langle PT(t)x, AT(t)x\rangle \\ &= \langle AT(t)x, PT(t)x\rangle + \langle PT(t)x, AT(t)x\rangle \\ &\leq -\|T(t)x\|^2,\end{aligned}$$

for all $x \in D(A)$ and $t \geq 0$. Which yields by integration over $(0, t)$ that

$$0 \leq F(t, x) \leq F(0, x) - \int_0^t \|T(s)x\|^2 ds.$$

Whence

$$\int_0^t \|T(s)x\|^2 ds \leq F(0, x) = \langle Px, x\rangle \quad \text{for all } t \geq 0 \text{ and } x \in D(A).$$

This inequality holds for all $x \in X$, by density. Moreover, since P is bounded then

$$\int_0^\infty \|T(s)x\|^2 ds \leq F(0, x) = \langle Px, x\rangle < \infty$$

for all $x \in X$. The exponential stability follows by Datko's theorem. ■

Now, we turn out to the stabilizability concept in the context of C_0-semigroups theory. First, we must fix some notations. For a strongly continuous semigroup $T(\cdot)$ with generator A on a Banach space $(X, \|\cdot\|)$, we denote by X_{-1} the completion of X with respect to the norm $\|x\|_{-1} := \|x\| + \|(\mu - A)^{-1}x\|$, for a fixed μ in the resolvent set $\rho(A)$. $T(\cdot)$ is extended to a semigroup $T_{-1}(\cdot)$ on X_{-1}. The generator A_{-1} of T_{-1} is an extension of A, precisely A is the part of A_{-1} on X. Moreover, we denote by X_1 the Banach space $D(A)$ endowed with the graph norm $\|x\|_1 := \|x\| + \|Ax\|$, $x \in D(A)$. Consider the abstract control system

$$\begin{aligned} \dot{z}(t) &= A_{-1}z(t) + Bu(t), \quad t \geq 0, \\ z(0) &= z^0 \in X. \end{aligned} \tag{ACS}$$

where A is the generator of a C_0-semigroup $T(\cdot)$ on a Banach space X, $B \in \mathcal{L}(U, X_{-1})$ for an other Banach space U. In general, the uncontrolled system, (ACS) with $u \equiv 0$, is not stable. To make it stable, one of the existing ideas in control theory is to apply a suitable input u. Because of simplicity for both implementation and analysis, the favourite tool to accomplish this objective is the use of a linear feedback relation of the form

$$u(t) = Fz(t). \tag{6}$$

Here the control $u(t)$ is determined as a linear function F of the current state $z(t)$. The problem is now reduced to find the feedback operator F such that the modified system,

$$\dot{z}(t) = (A_{-1} + BF)z(t), \quad z(0) = z^0.$$

realized by substituting the feedback law (6) into (ACS), is stable. That is, the operator

$$A^c := A_{-1} + BF \quad \text{with} \quad D(A^c) := \{x \in Y : (A_{-1} + BF)x \in X\}.$$

generates a stable semigroup. The system (ACS) is called the open-loop system while the modified system (ACS)-(6) is called the closed-loop system. To be more precise, *the open-loop system described by* (A, B) *is stabilizable if there exist* $X_1 \subseteq Y \subseteq X$ *and* $F \in \mathcal{L}(Y, U)$ *such that the feedback law* $u(t) = Fz(t)$ *is well defined in the sense of Weiss [28] and yields a stable closed-loop system.*

Recall that for finite dimensional systems, the (ACS) is stabilizable if and only if there exists a matrix F such that $A + BF$ is stable if and only if all of the eigenvalues of $A + BF$ lie on the open left half of the complex plan.

2.2 System Transformation

We start with simplifications that can be made to system (1). First, without loss of generality one can set $p \equiv 0$ since it can eliminated from equation (1) using the transformation

$$\tilde{y}(t,x) = y(t,x)e^{\int_0^x \frac{p(\sigma)}{2\tau(\sigma)}d\sigma}$$

with the compatibility changes of parameters

$$\tilde{q}(x) := q(x) - \frac{p'(x)}{2} - \frac{p^2(x)}{4\tau(x)}, \quad \tilde{u}(t) := e^{\int_0^1 \frac{p(s)}{2\tau(s)}ds}u(t).$$

Indeed, by explicit computations on can see that z satisfies (1) if and only if $\widetilde{z}$ satisfies (1) with the parameters $\tau,\ 0,\ \widetilde{q},\ \widetilde{u}$, instead of $\tau,\ p,\ q,\ u$.
Next, let $\Delta := \{(x,y):\ 0 \le y \le x \le 1\}$ and denote by H^i, $i = 0,1,2$, the usual Sobolev spaces on the interval $(0,1)$. In the next lemma, we recall from [20] a Volterra transformation which will be used to convert system (1) into an auxiliary one.

Lemma 2.3. *Let $k \in H^2(\Delta)$, and define the bounded operator $T_k : H^i \to H^i$, by*

$$(T_k\varphi)(x) := \varphi(x) - \int_0^x k(x,y)\varphi(y)dy. \qquad (T_k)$$

Then, T_k has a linear bounded inverse $T_k^{-1} : H^i \to H^i,\ i = 0,1,2$.

Assume that y satisfies (1) and set for $t \ge 0$, $x \in [0,1]$,

$$z(t,x) := (T_k y(t))(x) = y(t,x) - \int_0^x k(x,\sigma)y(t,\sigma)d\sigma. \qquad (7)$$

By integrating by parts from 0 to x, we get for $t > 0$,

$$\begin{aligned}\ddot{z}(t,x) &= \ddot{y}(t,x) - \int_0^x k(x,\sigma)\Big[[\tau(\sigma)y_\sigma(t,\sigma)]_\sigma + q(\sigma)y(t,\sigma)\Big]d\sigma \\ &= [\tau(x)y_x(t,x)]_x + q(x)y(t,x) - \tau(x)k(x,x)y_x(t,x) + \tau(0)k(x,0)y_x(t,0) \\ &+ \tau(x)k_\sigma(x,x)y(t,x) - \tau(0)k_\sigma(x,0)y(t,0) \\ &- \int_0^x \Big[[\tau(\sigma)k_\sigma(x,\sigma)]_\sigma + q(\sigma)k(x,\sigma)\Big]y(t,\sigma)d\sigma.\end{aligned}$$

Moreover,

$$
\begin{aligned}
[\tau z_x]_x =& [\tau(x)y_x(t,x)]_x - \tau(x)k(x,x)y_x(t,x) - [\tau(x)k(x,x)]_x y(t,x) \\
& - \tau(x)k_x(x,x)y(t,x) - \int_0^x [\tau(x)k_x(x,\sigma)]_x y(t,\sigma)d\sigma.
\end{aligned}
$$

Taking into account of the boundary condition at $x = 0$, we obtain

$$
\begin{aligned}
\ddot{z} - [\tau z_x]_x =& \Big[q(x) + 2\tau(x)\frac{d}{dx}(k(x,x)) + \tau'(x)k(x,x)\Big]y(t,x) \\
& - \int_0^x \Big[q(\sigma)k(x,\sigma) + \Big([\tau(\sigma)k_\sigma(x,\sigma)]_\sigma - [\tau(x)k_x(x,\sigma)]_x\Big)\Big]y(t,\sigma)d\sigma \\
& + \tau(0)k(x,0)y_x(t,0) - \tau(0)k_\sigma(x,0)y(t,0).
\end{aligned}
\tag{8}
$$

Then, $\ddot{z} - [\tau(x)z_x]_x = 0,$ in $(0,\infty)\times(0,1)$, if and only if the kernel k satisfies the PDE

$$
\begin{cases}
[\tau(x)k_x(x,\sigma)]_x - [\tau(\sigma)k_\sigma(x,\sigma)]_\sigma = q(\sigma)k(x,\sigma), & 0 \le \sigma \le x \le 1, \\
k_\sigma(x,0) = k(x,0) = 0, & 0 \le x \le 1, \\
2\tau(x)\frac{d}{dx}(k(x,x)) + \tau'(x)k(x,x) = -q(x), & 0 \le x \le 1.
\end{cases}
\tag{9}
$$

By solving the third differential equation of (9), we reach the following PDE governing the kernel k,

$$
\begin{cases}
[\tau(x)k_x(x,\sigma)]_x - [\tau(\sigma)k_\sigma(x,\sigma)]_\sigma = q(\sigma)k(x,\sigma), & 0 \le \sigma \le x \le 1, \\
k_\sigma(x,0) = k(x,0) = 0, & 0 \le x \le 1, \\
k(x,x) = -\frac{1}{\sqrt{\tau(x)}}\int_0^x \frac{q(s)}{2\sqrt{\tau(s)}}ds, & 0 \le x \le 1.
\end{cases}
\tag{10}
$$

We recall from [8] that, for any given functions $\tau \in C^2,\ q \in C^1$, the PDE (10) has a unique solution $k \in C^2(\Delta)$. Let now k be the solution of (10). In view of (8), the new state z satisfies

$$
\begin{cases}
\ddot{z}(t,x) = [\tau(x)z_x(t,x)]_x, & \text{in } (0,\infty)\times(0,1), \\
\ddot{z}(t,0) = \frac{\tau(0)}{M}z_x(t,0), & \text{in } (0,\infty), \\
z(t,1) = u(t) - \langle k(1,\cdot), y(t)\rangle_{L^2} =: v(t), & \text{in } (0,\infty), \\
z(0,x) = T_k y^0 =: z^0(x), \quad \dot{z}(0,x) = T_k y^1 =: z^1(x), & \text{in } (0,1),
\end{cases}
\tag{11}
$$

Remark 2.4. *We must point out that the perturbing term qy exerted from the entire domain in the initial system does not appear in the auxiliary one* (11)*. However, the boundary control at $x = 1$ is modified to $v(t) := u(t) - \big\langle k_1, y(t)\big\rangle_{L^2}$. Therefore, the Volterra transformation T_k, with a suitable kernel k, allows to convert the multiplicative perturbation, applied from the whole domain, into a boundary perturbation in the target system. This shows that the stabilization process for such hyperbolic PDEs can always be done from the boundaries. Further, the control design of the target system is usually much simpler than the initial one.*

2.3 Semigroups Representation of the Target System

To present the target system (11) as an abstract control system in the semigroups framework, we define $H_L^1 = \{f \in H^1 : f(1) = 0\}$, and we consider the Hilbert space $\mathcal{X} := H_L^1 \times L^2 \times \mathbb{C}$ endowed with the inner product[1]

$$\Big\langle (f, g, \xi), (\varphi, \psi, \eta)\Big\rangle := \int_0^1 [\tau f_x \overline{\varphi_x} + g\overline{\psi}]dx + M\xi\overline{\eta}.$$

Moreover, we introduce the operators

$$\begin{aligned}
&\mathcal{D}(A) := \big\{(f, g, \xi)^\top \in \mathcal{X} : f \in H^2,\ g \in H_L^1 : g(0) = \xi\big\},\\
&A\begin{pmatrix} f \\ g \\ \xi \end{pmatrix} = \begin{pmatrix} g \\ (\tau f')' \\ \tau(0)f'(0)/M \end{pmatrix}\\
&B\alpha = \alpha A_{-1}b,\ B \in \mathcal{L}(\mathbb{R}; \mathcal{X}_{-1}).
\end{aligned}$$

Then, by setting for $t \geq 0$,

$$X(t) = \begin{pmatrix} z - z(t,1)\mathbb{1} \\ \dot{z} \\ \dot{z}(t,0) \end{pmatrix}, \quad b = \begin{pmatrix} 0 \\ -\mathbb{1} \\ -1 \end{pmatrix}, \quad \text{with } \mathbb{1}(x) \equiv 1,$$

system (11) is reduced on $\mathcal{X}$ to an abstract control system;

$$\begin{aligned}
&\dot{X}(t) = A\Big(X(t) + \dot{v}(t)b\Big) = A_{-1}X(t) + B\dot{v}(t), \qquad t > 0,\\
&X(0) = (z^0 - z^0(1)\mathbb{1}, z^1, z^1(0))^\top =: X^0.
\end{aligned} \tag{12}$$

The following result confirms the well-posedness of the open-loop system of (12).

[1] $\overline{a}$ denotes the complex conjugate of the complex number a.

Theorem 2.5.
A is skew-adjoint. Therefore, A generates a group of isometries $(e^{tA})_{t\in\mathbb{R}}$ on $\mathcal{X}$.

Proof. Let $X = (f, g, \xi), Y = (\varphi, \psi, \eta)$ two elements of $\mathcal{D}(A)$. By integrating by parts we get

$$\begin{aligned}
\langle AX, Y\rangle &= \left\langle \begin{pmatrix} g \\ (\tau f')' \\ \frac{\tau(0)}{M} f'(0) \end{pmatrix}, \begin{pmatrix} \varphi \\ \psi \\ \eta \end{pmatrix} \right\rangle \\
&= \int_0^1 [\tau g'\overline{\varphi}' + (\tau f')'\overline{\psi}]\, dx + \tau(0) f'(0)\overline{\psi}(0) \\
&= \Big[\tau f'\overline{\psi} + \tau\overline{\varphi}' g\Big]_0^1 - \int_0^1 [\tau f'\overline{\psi}' + (\tau\overline{\varphi}')' g]\, dx + \tau(0) f'(0)\overline{\psi}(0) \\
&= -\int_0^1 [\tau f'\overline{\psi}' + (\tau\overline{\varphi}')' g]\, dx - \tau(0) g(0)\overline{\varphi}'(0) = -\langle X, AY\rangle.
\end{aligned}$$

It follows that $\mathcal{D}(A) \subset \mathcal{D}(A^*)$ and $A^* v = -Av$ for all $v \in \mathcal{D}(A)$. To show that $\mathcal{D}(A) = \mathcal{D}(A^*)$, we shall prove that $0 \in \rho(A) \cap \rho(A^*)$ and $R(0, A)$ is skew adjoint, i.e, $R(0, A^*) = R(0, A)^* = -R(0, A)$. In fact, let $Y = (\varphi, \psi, \eta)^\top \in \mathcal{X}$. The equation,

$$AX = Y, \;\; X = (f, g, \xi)^\top \in \mathcal{D}(A),$$

which is equivalent to

$$\begin{cases} g = \varphi \\ f \in H^2, \\ (\tau f')' = \psi \\ f(1) = 0, \;\; \tau(0) f'(0) = M\eta \end{cases}$$

is uniquely solvable by the vector

$$v_0 = \begin{pmatrix} f : x \mapsto -\int_x^1 \frac{M\eta}{\tau(s)} ds - \int_x^1 \int_0^s \frac{\psi(\sigma)}{\tau(s)} d\sigma ds \\ g = \varphi \\ \xi = \varphi(0) \end{pmatrix}$$

Therefore, $0 \in \rho(A)$. On the other hand, due to the inclusion $\mathcal{D}(A) \subset \mathcal{D}(A^*)$, the vector $(-v_0)$ also solves the equation

$$A^* X = Y, \;\; X \in \mathcal{D}(A^*). \tag{13}$$

Moreover, since $0 \in \rho(A)$, then $A\mathcal{D}(A) = \mathcal{X}$. Which implies that, if $\langle A^* X, v\rangle = \langle X, Av\rangle = 0$ for all $v \in \mathcal{D}(A)$ then $X \equiv 0$. This shows the injectivity of A^* and hence the uniqueness of the solution of (13). Consequently, $0 \in \rho(A^*)$ and $v_0 = R(0, A)Y = -R(0, A^*)Y$ for all $Y \in \mathcal{X}$. Therefore, $\mathcal{D}(A^*) = R(0, A^*)\mathcal{X} = R(0, A)\mathcal{X} = \mathcal{D}(A)$. The rest of the assertion is a well-known Ston's result of skew-adjoint operator on a Hilbert space, see e.g. [11, Thm 3.24]. ■

In the next lemma, we summarize the useful properties of B^*, the adjoint of B. The proof follows from explicit computations, so we omit it.

Lemma 2.6.
(i) $B^* \in \mathcal{L}((\mathcal{X}_{-1})^*, \mathbb{R}), \quad \mathcal{D}(A) \subseteq \mathcal{D}(B^*) = (\mathcal{X}_{-1})^* \subseteq \mathcal{X}$,
(ii) $B^*{}_{|\mathcal{D}(A)} = -b^* A$,
(iii) $B^* X = \tau(1) f'(1), \mathcal{D}(B^*) = \{X = (f, h, \xi) \in \mathcal{X} : f' \text{ is continuous at } 1\}$,
(iv) $b \in \mathcal{D}(B^*)$, *and* $B^* b = 0$.

3. Exponential Stabilization of the Target System

The aim of this section is to construct a controller v which stabilizes the auxiliary system (11). To this end, it is convenient to introduce the functional,

$$E(t) = \frac{1}{2}\left\{\int_0^1 \left[\tau(x)|z_x(t,x)|^2 + |\dot{z}(t,x)|^2\right]dx + M|\dot{z}(t,0)|^2\right\}, \tag{14}$$

where $z(t, x)$ satisfy (11). The time derivative of E gives

$$\dot{E}(t) = \frac{1}{2}\tau(1)\left(z_x(t,1)\overline{\dot{v}(t)} + \overline{z_x(t,1)}\dot{v}(t)\right) \tag{15}$$

Choosing in (15) the feedback law,

$$\dot{v}(t) = -\gamma\tau(1)z_x(t,1), \qquad \gamma > 0, \tag{16}$$

we obtain $\dot{E}(t) = -\gamma\left|\tau(1)z_x(t,1)\right|^2$. This implies that the boundary feedback law (16) forces E to decreases with time t. Next, we will show the exponential decay of E. To do so, let us set $X(t) = (z - z(t,1)\mathbb{1};\ \dot{z};\ \dot{z}(t,0))^\top$, for $t \geq 0$.

Then, the boundary feedback law (16) will be $\dot{v}(t) = -\gamma B' X(t)$ and the closed-loop system (11), (16) is reduced to

$$\dot{X} = A(X - \gamma b B^* X) =: A^c X, \qquad X(0) = X^0, \quad t > 0, \tag{17}$$
$$\mathcal{D}(A^c) = \{X \in \mathcal{D}(B^*); \;\; X - \gamma b B^* X \in \mathcal{D}(A)\}.$$

Moreover, A^c has the explicit form

$$A^c \begin{pmatrix} f \\ g \\ \xi \end{pmatrix} = \begin{pmatrix} g + \gamma\tau(1) f'(1)\mathbb{1} \\ (\tau f')' \\ \tau(0) f'(0)/M \end{pmatrix},$$
$$\mathcal{D}(A^c) := \{(f, g, \xi) \in (H^1_L \cap H^2) \times H^1 \times \mathbb{R} : g(0) = \xi, g(1) = -\gamma\tau(1) f'(1)\}.$$

The scheme depicted in figure 1 describes, in a simplified way, the structure of the closed loop (11)-(16) which is represented by the evolution equation (17). The next theorem will confirm the well-posedness of the Cauchy problem (17).

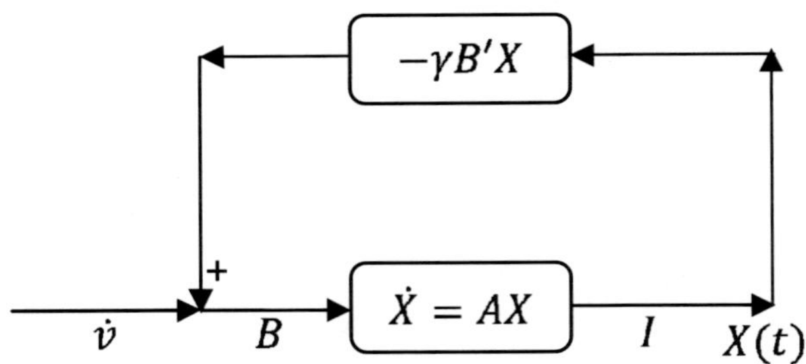

Figure 1. The closed-loop system (11)-(16), represented by (17), corresponds to the evolution operator A, control operator B (input $\dot{v}$), observation operator I (ouput X) and feedback law $\dot{v} := -\gamma B^* X$.

Theorem 3.1.
(i) $\Re\langle A^c X, X\rangle = -\gamma |B^* X|^2$, *for all* $X \in \mathcal{D}(A^c)$.
(ii) $0 \in \rho(A^c)$.
(iii) A^c *generates on* $\mathcal{X}$ *a contraction* C^0*-semigroup* $(e^{tA^c})_{t\geq 0}$.

Proof. (i) Let $X \in \mathcal{D}(A^c)$,

$$\begin{aligned} \langle A^c X, X\rangle &= \langle AX - \gamma A_{-1} b B^* X, X\rangle \\ &= \langle AX, X\rangle - \gamma\langle A_{-1} b B^* X, X\rangle \\ &= \langle AX, X\rangle - \gamma\langle B^* X, B^* X\rangle. \end{aligned}$$

Similarly, $\langle X, A^c X\rangle = \langle X, AX\rangle - \gamma\langle B^*X, B^*X\rangle$. Thus, the result (i) follows readily from the fact that A is skew-adjoint, due to Theorem 2.5.
(ii) For $Y \in \mathcal{X}$ the equation $A^c X = Y,\ X \in \mathcal{D}(A^c)$ is equivalent to

$$X = A^{-1}Y + \gamma b B^* X \tag{18}$$

If such a solution exists it may belong to $\mathcal{D}(A^c)$ (which is included in $\mathcal{D}(B^*)$). Taking into account of the fact that $B^*b = 0$ and $B^*{}_{|\mathcal{D}(A)}$ is invertible, due to Lemma 2.6, the solution X of (18) exists if and only if

$$B^*X = B^*A^{-1}Y = -b^*AA^{-1}Y = -b^*Y.$$

From (18), it follows that

$$X = A^{-1}Y - \gamma bb^*Y = (A^{-1} - \gamma bb^*)Y,$$

which belongs to $\mathcal{D}(A^c)$. This shows that A^c is invertible and $(A^c)^{-1} = A^{-1} - \gamma bb^*$.
(iii) It results from (i) that A^c is $0-$dissipative. Moreover, due to (ii), A^c is surjective. This suffices to assert that A^c generates a C^0-semigroup of contractions. ∎

For the purpose of proving the exponential stability of the closed-loop system (11),(16), we introduce on $\mathcal{X}$ the operators

$$N\begin{pmatrix} f \\ g \\ \xi \end{pmatrix} = \begin{pmatrix} -\int_x^1 \frac{\mu(s)g(s)}{\tau(s)}ds \\ \mu f' \\ 0 \end{pmatrix} \tag{19}$$

$$Q_\varepsilon = \varepsilon I + N,$$

with a precise function μ and a sufficiently large ε. As we will see in the proof of the next lemma, the function μ should satisfy the differential equation

$$\mu'\tau - \mu\tau' = \tau, \quad \mu(0) = M, \tag{20}$$

namely $\mu(x) = \tau(x)\left[\frac{M}{\tau(0)} + \int_0^1 \frac{ds}{\tau(s)}\right]$. Moreover, ε should be greater than the constant ε_0,

$$\varepsilon_0 := \max\left(\delta := \frac{\mu(1)}{\sqrt{\tau(0)}},\ \frac{\gamma^2\tau(1)^2\mu(1) + \mu(1)\tau(1)}{2\gamma\tau(1)^2} =: \kappa\right) \tag{21}$$

In what follows we denote by $\|\cdot\|$ the norm on $\mathcal{X}$ associated to the inner product. Moreover, we briefly write Q instead of Q_ε when the notation does not lead to any confusion.

Lemma 3.2. *For $\varepsilon > \varepsilon_0$, the following assertions hold,*
(i) $N \in \mathcal{L}(\mathcal{X})$, with $\|N\|_{\mathcal{L}(\mathcal{X})} \leq \delta$ and $N^* = N$,
(ii) $Q^* = Q$,
(iii) Q is positive and invertible,

Proof. (i)-(ii) follows from explicit computations.
(iii) The positivity of Q is due to

$$Q = \varepsilon I + N \geq \left(\varepsilon - \|N\|_{\mathcal{L}(\mathcal{X})}\right) I \geq (\varepsilon - \delta)\, I \geq 0.$$

On the other hand, $Q = \varepsilon I + N = \varepsilon(I + N/\varepsilon)$, is invertible since $\|\frac{N}{\varepsilon}\|_{\mathcal{L}(\mathcal{X})} \leq \frac{\varepsilon_0}{\varepsilon} < 1$.

■

Now we are in a comfortable position to deal with the exponential stability of the closed-loop system (11),(16), by which we refer to the exponential stability of the semigroup $(e^{tA^c})_{t\geq 0}$.

Theorem 3.3.
(i) $\langle A^c X, QX\rangle + \langle X, QA^c X\rangle \leq -\|X\|^2$, for all $X \in \mathcal{D}(A^c)$.
(ii) The semigroup $(e^{tA^c})_{t\geq 0}$ is exponentially stable. Moreover, the estimate

$$\|e^{tA^c}\| \leq \sqrt{\frac{\varepsilon + \delta}{\varepsilon - \delta}}\; e^{-\frac{1}{2(\varepsilon+\delta)}t}, \quad t \geq 0, \tag{22}$$

holds for $\varepsilon > \varepsilon_0$.
In particular, the feedback law (16) *exponentially stabilizes* (11).

Proof. (i) Let $X = (f, g, \xi) \in \mathcal{D}(A^c)$. By using (20) and the fact that $\mu' \geq 1$, we have

$$\begin{aligned}
\langle A^c X, NX\rangle + \langle NX, A^c X\rangle &= \int_0^1 \mu(g'\overline{g} + \overline{g'}g) + \mu\tau(f''\overline{f'} + \overline{f''}f') + 2\mu\tau'|f'|^2 dx \\
&= \left[\mu|g|^2 + \mu\tau|f'|^2\right]_0^1 - \int_0^1 \mu'|g|^2 + \left[\tau\mu' - \tau'\mu\right]|f'|^2 dx \\
&\leq -\|X\|^2 + \mu(1)\left(|g(1)|^2 + \tau(1)|f'(1)|^2\right) \\
&\leq -\|X\|^2 + \mu(1)\left(\gamma^2\tau(1)^2 + \tau(1)\right)|f'(1)|^2.
\end{aligned}$$

Which implies, by Theorem 3.1-(i), that

$$\begin{aligned}\langle A^c X, QX\rangle + \langle QX, A^c X\rangle &= -2\gamma\varepsilon |B^* X|^2 + 2\Re\langle A^c X, NX\rangle \\ &= -2\gamma\varepsilon |\tau(1) f'(1)|^2 + 2\Re\langle A^c X, NX\rangle \\ &\leq -\|X\|^2 + \left[\gamma^2 \tau(1)^2 \mu(1) + \tau(1)\mu(1) - 2\gamma\varepsilon\tau(1)^2\right] |f'(1)|^2 \\ &\leq -\|X\|^2 + (\kappa - \varepsilon) 2\gamma\tau(1)^2 |f'(1)|^2\end{aligned}$$

This proves inequality (i) since $\kappa - \varepsilon < 0$.

(ii)The statement (i) means that the Lyapunov operator inequality (5) has a solution $K = K^* \in \mathcal{L}(\mathcal{X})$, namely $K = Q_\varepsilon$ for $\varepsilon > \varepsilon_0$. The exponential stability of $(e^{tA^c})_{t\geq 0}$ follows from Proposition 2.2. To shows estimate (22), let $\varepsilon > \varepsilon_0$. The corresponding operator Q_ε (simplified by Q) is positive and auto-adjoint. Therefore, the operator $Q^{\frac{1}{2}}$ is well defined, as the unique solution of the equation $R^2 = Q,\ R \in \mathcal{L}(\mathcal{X})$. Moreover, $Q^{\frac{1}{2}}$ is positive and auto adjoint too. This enables us to define on $\mathcal{X}$ the new norm

$$\|X\|_Q^2 := \left\langle Q^{\frac{1}{2}} X, Q^{\frac{1}{2}} X \right\rangle = \langle X, QX\rangle .$$

which is equivalent to $\|\cdot\|$ and it is not difficult to see the estimate

$$\sqrt{\varepsilon - \delta}\,\|X\| \leq \|X\|_Q \leq \sqrt{\varepsilon + \delta}\,\|X\|. \tag{23}$$

In fact, let $X(t) = e^{tA^c} X^0$ be the solution of the evolution equation (17) starting from $X^0 \in \mathcal{D}(A^c)$ and define for $t \geq 0$,

$$V(t) = \frac{1}{2}\|X(t)\|_Q^2.$$

It follows from Lemma 3.2-(iv) and estimate (23),

$$\dot{V}(t) = \Re\langle A^c X(t), QX(t)\rangle \leq -\frac{1}{2}\|X(t)\|^2 \leq -\frac{1}{(\varepsilon+\delta)} V(t), \ \ t \geq 0.$$

Which implies that $V(t) \leq e^{-\frac{t}{(\varepsilon+\delta)}} V(0)$ and by (23) we have

$$\|X(t)\|^2 \leq \frac{\varepsilon + \delta}{\varepsilon - \delta}\, e^{-\frac{t}{(\varepsilon+\delta)}} \|X(0)\|^2, \ \ t \geq 0.$$

Therefore,

$$\|e^{tA^c} X^0\| \leq \sqrt{\frac{\varepsilon+\delta}{\varepsilon-\delta}}\, e^{-\frac{t}{2(\varepsilon+\delta)}} \|X^0\|, \ \ t \geq 0,$$

for all $X^0 \in \mathcal{D}(A)$, hence for all $X^0 \in \mathcal{X}$ by density. ■

Theorem 3.3 shows that the controller (16) exponentially stabilizes system (11) at the decay rate $r = -\frac{1}{2(\varepsilon+\delta)}$. Moreover, the estimate (22) gives a representation of the decay rate r by means of the system parameters ε, δ.

Remark 3.4.
Observe that

$$\kappa - \delta = \frac{1}{2\gamma\tau(1)\sqrt{\tau(0)}}(\gamma^+ - \gamma^-)(\gamma^+ + \gamma^-)$$

where $\gamma^- := \frac{1}{\tau(0)} - \sqrt{\frac{1}{\tau(0)} - \frac{1}{\tau(1)}}$, $\gamma^+ := \frac{1}{\tau(0)} + \sqrt{\frac{1}{\tau(0)} - \frac{1}{\tau(1)}}$,
Hence,

$$\begin{array}{ll} \varepsilon_0 = \delta, & \text{if } \gamma \in [\gamma^-, \gamma^+], \text{ and} \\ \varepsilon_0 = \kappa, & \text{if } \gamma \in]0, \gamma^-[\cup]\gamma^+, +\infty[. \end{array}$$

Therefore, in order to accelerate the stabilization process, one has to choose by estimate (22), ε *closer to* ε_0. *Which corresponds to choose*

$$\begin{array}{ll} \varepsilon \text{ closer to } \delta, & \text{if } \gamma \in [\gamma^-, \gamma^+], \\ \varepsilon \text{ closer to } \kappa, & \text{if } \gamma \in]0, \gamma^-[\cup]\gamma^+, +\infty[. \end{array}$$

4. Exponential Stabilization of the Initial System

With the help of the isomorphism T_k and the exponential stability of the target system (11) we will infer the EXS of system (1). For this purpose, we introduce the auxiliary operator

$$\mathcal{P}_k \begin{pmatrix} f \\ g \\ \xi \end{pmatrix} = \begin{pmatrix} \varphi - \varphi(1)\mathbb{1} =: Pf \\ \psi \\ \xi \end{pmatrix}, \qquad \varphi = T_k f, \ \psi = T_k g,$$

where k is the solution of the kernel PDE (10). To state useful properties of $\mathcal{P}_k$ we set

$$h_k := T_k^{-1}\mathbb{1}, \quad \overline{k} = h_k(1) \quad \text{and} \quad \mathbb{1}_k := h_k/\overline{k}, \quad \text{if } \overline{k} \neq 0.$$

The following lemma is proved using direct computations.

Lemma 4.1. *If* $\overline{k} \neq 0$ *then*
(i) P *has a bounded inverse* P^{-1} *on* H^1_L *given by*

$$P^{-1}\varphi = f - f(1)\mathbb{1}_k, \quad \text{for } \varphi \in H^1_L,$$

where $f = T_k^{-1}\varphi$.
(ii) $\mathcal{P}_k$ *is an isomorphism on* $\mathcal{X}$ *with* $\mathcal{P}_k^{-1} = (P^{-1}, T_k^{-1}, I)$

For $\overline{k} = 0$ the operator P, hence $\mathcal{P}_k$, is not invertible. In fact, if $\overline{k} = 0$, the function h_k will belong to H_L^1 and $Ph_k = 0$. However, if P is invertible then $h_k = 0$, which leads to the contradiction $\mathbb{1} = T_k h_k = 0$.
On the other hand, due to $T_k h_k = \mathbb{1}$, the function h_k satisfies the PDE

$$(\tau h_k')' + qh_k = 0, \quad h_k(0) = 1, \quad h_k'(0) = 0.$$

Therefore, the function h_k depends on τ, q, hence $\overline{k} = \overline{k}(\tau, q)$. However such functions for which $\overline{k}(\tau, q) = 0$ are rather rare.
Now, let $X(t) = (z(t) - z(t,1)\mathbb{1}, \dot{z}(t), \dot{z}(t,0))$ be the solution of the evolution equation (17), representing the closed-loop (11), (16). Substituting (16) in the 1-end boundary condition of (1) we get for $t > 0$,

$$\begin{aligned} \dot{u}(t) &= -\gamma\tau(1)B^*X(t) + \langle k_1, \dot{y}(t)\rangle_{L^2} \\ &= -\gamma\tau(1)z_x(t,1) + \langle k_1, \dot{y}(t)\rangle_{L^2}, \end{aligned} \tag{24}$$

where $k_1(\sigma) := k(1,\sigma)$. For a function y satisfying (1), let $Y(t) = (y(t) - y(1)\mathbb{1}_k, \dot{y}, \dot{y}(0))$, and consider the operator

$$A_q^c \begin{pmatrix} f \\ g \\ \xi \end{pmatrix} = \begin{pmatrix} g - g(1)\mathbb{1}_k \\ (\tau f')' + qf \\ \tau(0)f'(0)/M \end{pmatrix},$$

$$\begin{aligned} \mathcal{D}(A_q^c) :=&\{(f, g, \xi)^\top \in H_L^1 \cap H^2 \times H^1 \times \mathbb{R} : g(0) = \xi, \\ & g(1) = -\gamma\tau(1)(T_k f)'(1) + \langle k_0, g\rangle_{L^2}\}. \end{aligned}$$

Then system (1),(24) is reduced to the evolution equation

$$\dot{Y}(t) = A_q^c Y(t), \quad t > 0, \quad Y(0) = Y^0, \tag{25}$$

where $Y^0 = (y^0 - y^0(1)\mathbb{1}_k, y^1, y^1(0))^\top$. It will useful for the rest to prove the following results ;

Theorem 4.2. *Assume that* $\overline{k} \neq 0$. *Then,*
(i) $A_q^c = \mathcal{P}_k^{-1} A^c \mathcal{P}_k$,
(ii) A_q^c *generates a* C_0*-semigroup* $(e^{tA_q^c})_{t\geq 0}$ *satisfying*

$$e^{tA_q^c} = \mathcal{P}_k^{-1}\, e^{tA^c}\, \mathcal{P}_k, \quad t \geq 0. \tag{26}$$

Consequently, system (1)-(24) *is well-posed.*

Proof. (i) Let $v = (f, g, \xi) \in \mathcal{D}(A_q^c)$ and set $\varphi = T_k f,\ \psi = T_k g$. Then $\mathcal{P}_k v \in \mathcal{D}(A^c)$ and

$$A^c \mathcal{P}_k v = A^c \begin{pmatrix} \varphi - \varphi(1)\mathbb{1} \\ \psi \\ \xi \end{pmatrix} = \begin{pmatrix} \psi + \gamma\tau(1)\varphi'(1)\mathbb{1} \\ (\tau\varphi')' \\ \tau(0)\varphi'(0)/M \end{pmatrix}$$

On the other hand,

$$\mathcal{P}_k A_q^c v = \mathcal{P}_k \begin{pmatrix} g - g(1)\mathbb{1}_k \\ (\tau f')' + qf \\ \tau(0)f'(0)/M \end{pmatrix} = \begin{pmatrix} P(g - g(1)\mathbb{1}_k) \\ T_k[(\tau f')' + qf] \\ \tau(0)\varphi'(0)/M \end{pmatrix}$$

By construction of the kernel PDE (10), see Section (2.2), we have

$$T_k[(\tau f')' + qf] = \tau(\varphi')'$$

In addition

$$P(g-g(1)\mathbb{1}_k) = \psi - \frac{g(1)}{k}\mathbb{1} - \left(\psi(1) - \frac{g(1)}{k}\right)\mathbb{1} = \psi - \psi(1)\mathbb{1} = \psi + \gamma\tau(1)\varphi'(1)\mathbb{1}$$

This shows that $\mathcal{P}_k A_q^c v = A^c \mathcal{P}_k v$ for all $v \in \mathcal{D}(A)$.
(ii) It yields from (i) that A_q^c is isomorphic (similar) to A^c, then by standard semigroup arguments A_q^c generates a C_0-semigroup $(e^{tA_q^c})$ which is also similar to (e^{tA^c}) in the sense that it satisfies (26), see e.g [11, p.59]. ■

Theorem 4.2 establishes the relationship between the solution Y of the closed-loop (1), (24) starting from Y^0, represented by (25) and the solution X of the target closed-loop (11), (16) starting from $X^0 = \mathcal{P}_k Y^0$, described by (17). Indeed, due to (26), we have

$$X(t) = \mathcal{P}_k Y(t), \quad t \geq 0.$$

This allows to rewrite the boundary feedback law (24) by means of Y,

$$\dot{u}(t) = -\gamma\tau(1)B^*\mathcal{P}_k Y(t) + \langle k_1, \dot{y}(t)\rangle_{L^2}, \qquad t > 0. \tag{27}$$

On the other hand, the x-derivative of z from (7) one gets for $x = 1$,

$$z_x(t,1) = y_x(t,1) - k_1(1)u(t) - \langle k_2, y\rangle_{L^2}, \qquad t > 0,$$

where $k_2(\sigma) := k_x(1, \sigma)$. Substituting (4) in (24), we obtain the boundary feedback law (24) in terms of the solution y of the original system,

$$\dot{u}(t) = -\gamma\tau(1)\left[y_x(t,1) - k_1(1)u(t) - \langle k_2, y(t)\rangle_{L^2}\right] + \langle k_1, \dot{y}(t)\rangle_{L^2},\ t > 0. \quad (28)$$

The explicit solution of (28) with the initial data $u(0) = 0$ is

$$u(t) = -e^{\gamma\tau(1)k_1(1)t}\int_0^t \left[\gamma\tau(1)y_x(s,1) - \langle \gamma\tau(1)k_2, y(s)\rangle_{L^2} - \langle k_1, \dot{y}(s)\rangle_{L^2}\right]ds, \quad (29)$$

for $t \geq 0$. Now we are ready to state the second main result of this chapter.

Theorem 4.3. *If $\overline{k} \neq 0$ then the semigroup $(e^{tA_q^c})_{t\geq 0}$ is exponentially stable. Moreover, for $\varepsilon > \varepsilon_0$, there exists a constant $C > 0$ such that*

$$\|e^{tA^c}\| \leq C\sqrt{\frac{\varepsilon+\delta}{\varepsilon-\delta}}\, e^{-\frac{t}{2(\varepsilon+\delta)}}, \quad t \geq 0.$$

In particular, the controller (29) exponentially stabilizes system (1) at the decay rate $r = \frac{1}{2(\varepsilon+\delta)}$.

Proof. The proof is a direct consequence of expression (26), the exponentially stability of $(e^{tA^c})_{t\geq 0}$ established in Theorem 3.3 and the boundedness of the isomorphism $\mathcal{P}_k$, due to Lemma 4.1. ∎

We must mention here that the control design presented at this level is given for $p \equiv 0$. However, when the zero set of p is (Lebesgue) negligible, the exponential stabilizer $\widetilde{u}$ of system (1) will be in view of Section 2.2,

$$\tilde{u}(t) := e^{\int_0^1 \frac{p(s)}{2\tau(s)}ds}u(t)$$

where u is given by (29).

Conclusion

The main idea of studying the exponentially stabilizability of the hyperbolic equation (1) lies on the fact that the (exponentially) stability property of a given system does not change when the system is subject to a similarity (isomorphism). So, we first convert the original system (1) into a desired target one by using a Volterra transformation T_k, with a kernel k chosen to satisfy an appropriate PDE. Second, We show that the target system is exponentially stable in semigroup setting by providing an explicit solution of the Lyapunov's equation (4). The exponentially stability of the initial system is derived using the inverse of an auxiliary isomorphism $\mathcal{P}_k$ (given in terms of T_k). More precisely, the associated semigroups are similar. A first advantage of the proposed method is that it transforms a perturbation from the whole domain into a perturbation from the boundary of control. Second, it gives information about the controller construction by means of a well-known auxiliary system. Third, it provides an explicit expression of the rate of decay by means of the system parameters and the solution of the kernel PDE.

It should be stressed that the Volterra transformation T_k, with adequate kernel k, converts a large class of PDEs without effects on their topological properties. Therefore, various properties of parabolic and hyperbolic PDEs, namely optimizability, controllability, observability,. . . can be treated by using similar techniques of the adopted strategy.
According to the pursued strategy, the treatment of system (1) in time-depending setting, i.e., when the coefficients p, q, τ depend on time t, requires to solve a time depending PDE of a kernel $k(t, x, \sigma)$ and to investigate the well-posedness of a non-autonomous evolution equation representing the closed-loop system. A promising way to treat the case of time depending $p(t), q(t), \tau(t)$ may be the frozen coefficient method and integral product as restored in recent researches [12], [23] and [24] or the interesting survey on recent progress concerning maximal regularity of non-autonomous equations governed by time-dependent forms on a Hilbert spaces summarized by W. Arendt, S. Fackler and D. Dier in [2].

References

[1] D'Andréa-Novel, B., Boustany, F., Conrad, F. & B. P. Rao, Feedback stabilization of a hybrid PDE-ODE system: application to an overhead crane, *Math. Control Signals Systems*, Vol. 17, No. 1, pp. 1-22, 1994.

[2] Arendt, W Fackler, S and Dier, D. *J. L. Lions problem on maximal regularity,* availaible on arXiv:1612.03676v1 [math.FA] 12 Dec 2016.

[3] D'Andréa-Novel, B., Coron, J. M. Exponential stabilization of an overhead crane with flexible cable via a cascade approach, *Automatica*, Vol.36, N. 4, pp 587-593, 2000.

[4] Bardos, C., Lebeau, G.and Rauch, J. Sharp sufficient conditions for the observation, control, and stabilization of waves from the boundary, *SIAM J. Control. Optim.*, 30 (1992), pp. 1024-1065.

[5] Cerpa, E., Krstic, M. and Smyshlayev, M., Boundary Stabilization of 1-D Wave Equation with In-domain Antidamping, *SIAM J. Control Optim.* Vol 48, N. 6, pp. 4014–4031, 2010.

[6] Conrad, F., and Mifdal, A. Strong stability of a model of an overhead crane, *Control and Cybernetics*, Vol. 27, No. 3, 1998.

[7] Datko,R. Extending a theorem of A.M. Liapunov to Hilbert space, *J. Math. Anal. Appl.* 32 (1970), 610-616.

[8] Elharfi, A. Explicit construction of a boundary feedback law to stabilize a class of parabolic equations, *Differential and Integral Equations*, Vol. 21, N. 3, pp. 351-362, 2008.

[9] Elharfi, A. Control design of an overhead crane system from the perspective of stabilizing undesired oscillations, *IMA Journal of Mathematical Control and Information* 28, 267-278, (2011).

[10] Elharfi, A. Exponential Stabilization of a Class of 1-D hyperbolic PDEs, *J. Evol. Equ.* 16 (2016), 665-679.

[11] Engel K., Nagel, R. *One-parameter Semigroups for Linear Evolution Equations*, Springer–Verlag 2000.

[12] El-Mennaoui, O. Keyantuo, V. and Laasri, H Infinitesimal product of semigroups, *Ulmer Seminare. Heft* 16 (2011), 219–230.

[13] Gearhart, L. Spectral theory for contraction semigroups on Hilbert spaces, *Trans. Amer. Math. Soc.* 236 (1978), 385-394.

[14] Grabowski, P. The motion planning problem and exponential stabilization of a heavy chain. Part I, *International Journal of Control*, vol. 82, N. 8, pp. 1539-1563, 2009

[15] Grabowski, P.*Abstract semigroup model of heavy chain system with application to a motion planning problem,* Proceedings of 9th IEEE International Conference: Methods and Models in Automation and Robotics 25-28 August, Miedzyzdroje, Poland, 2003, pp. 77 - 86.

[16] Hille, E., Phillips, R.S. Functional Analysis and Semigroups, *Amer. Math. Soc. Coll. Publ.*, vol. 31, Amer. Math. Soc., 1957.

[17] V. Komornik and P. Loreti. Fourier Series in Control Theory, *Springer Monogr. Math.*, Springer-Verlag, New York, 2005.

[18] Lasiecka. I., and Triggiani, R. Riccati equations for hyperbolic partial diferential equations with $L2(0,T;L2(\Gamma))$-Dirichlet boundary terms, *SIAM J.Control Optim.*, 24, pp. 884-925, 1986.

[19] Lions, J.L. *Contrôlabilité exacte, perturbations et stabilisation des ystèmes distribués*, Masson, Paris, 1988.

[20] Liu, W. Boundary feedback stabilization for an unstable heat equation, *SIAM J. Control Optim.* Vol 42, N. 3, pp. 1033–1043, 2003.

[21] Littman, W. A generalization of a theorem of Datko and Pazy,: *Lecture Notes in Control and Inform. Sci.* 130, Springer-Verlag, Berlin (1989) 318-323.

[22] Pazy, A. Semigroups of Linear Operators and Applications to Partial Differential Equations, *Appl. Math. Sci.*, vol. 44, Springer-Verlag, 1983.

[23] Sani, A and Laasri, H. Evolution equations governed by Lipschitz continuous non autonomous forms, *Czechoslovak Mathematical Journal*, 65 (140) (2015), 475-491.

[24] Sani, A *Intégrale produit et régularité maximale du problème de Cauchy non autonome*, Phd Thesis, Ibn Zohr University, Agadir, Morocco.

[25] Rao, B. *Exponential stabilization of a hybrid system by dissipative boundary damping,* Proceedings of the Second European Control Conference, Groningen, pp. 314-317.

[26] D.L.Russell, Controllability and stabilizability theory for linear partial differential equations: Recent progress and open questions, *SIAM J. Control Optim.* Vol 20, pp. 639-739, 1978.

[27] Tucsnak, M. Boundary stabilization for the stretched string equation, *Differential Integral Equations,* Vol. 6, pp. 925-935, 1993.

[28] Weiss, G., Reberber, R. Optimizability and estimability for infinite-dimensional systems. *SIAM J. Control optim.* Vol 39, pp 1204-1232, 2000.

[29] Weiss, G. Weak L^p-stability of a linear semigroup on a Hilbert space implies exponential stability, *J. Diff. Eq.* 1988, 269-285.

INDEX

F

G

H

I

K

L

M

N

O